Stalk and Kill

Adrian Gilbert

Stalk and Kill

THE SNIPER EXPERIENCE

St. Martin's Press
New York

The author and publishers acknowledge with thanks permission to use the
extract from *Homage to Catalonia* by George Orwell, which appears on pages
235–8, © the estate of the late Sonia Brownell Orwell and Martin Secker &
Warburg Ltd.

ISBN 0-312-17030-0

Contents

Contents

Preface

Sniping in the armed forces of the West has had a difficult history. In peace, snipers tended to be considered an irrelevance. In war, the value of sniping was eventually accepted, and time and resources were channelled into its development. But on the resumption of peace, sniping again fell into the doldrums. This process was repeated time after time, and only in the 1970s and 1980s was the cycle finally broken, so that snipers were organized and trained on a peacetime footing to be ready for war.

This book traces the course of sniping's stop–go history through the eyes of the sniper. First-hand accounts from the men in the field provide a revealing insight into this lonely and dangerous occupation. The special pressures faced by a sniper – not least being the sight of the man he is about to kill – have imposed great physical and psychological strains, from which only the toughest survive. And if weapons and tactics have developed over the years, then the human factor remains constant. The sniper is the ultimate hunter in a game where the quarry shoots back. This hunting instinct has changed little over the last two centuries, linking the sharpshooter of the American Revolution to the professionally trained sniper of today.

The eyewitness accounts used in this book are reproduced verbatim whenever possible, but obvious errors have been corrected and minor confusions clarified. It has not been thought necessary to indicate where quotations have been abridged.

Acknowledgements

The Imperial War Museum was a vital source of information, and my thanks go to the staff of the Departments of Printed Books, Documents, Sound Records and Photographs. My appreciation also goes to Howard Mitchell, the librarian at the Ministry of Defence Pattern Room, and to the staff of the British Library, who gave me access to many obscure volumes. Material on American subjects was provided by the US Army Military History Institute at Carlisle, Pennsylvania, and the US Marine Corps Museum at the Navy Yard, Washington, DC.

Of the many individuals who helped with this book, a special mention must go to Harry M. Furness. A sniper with a distinguished service record from north-west Europe during 1944–45, he was unfailing in providing me with encouragement and information, as well as several firsthand accounts of his sniping experiences (my thanks here extend to Doug Scott, the archivist of Mr Furness's battalion, the York and Lancaster Regiment). Eyewitness material and other information was kindly supplied to me by the following individuals: Keith Cook, David Cooper, Mick Harrison, Skippy Hampstead, Mike James, Lee Marvin, Bob Morrison, Tom Nowell, Roger Payne, Mark Spicer and Richard Russell. A number of other contributors have not been named for security reasons, but to them I also include my thanks.

Nigel Greenaway is conducting ongoing research into British sniping and was kind enough to supply me with two firsthand accounts. Any sniper or interested individual who would be prepared to help him with his research can contact him via Adrian Gilbert and the publishers.

In the United States I had the good fortune to receive first-rate assistance from Roy and Norman Chandler, who produce the *Death from Afar: Marine Corps Sniping* series. Roy Chandler generously gave me permission to use *Death from Afar* material as well as supplying me with excellent leads. One of these was Thomas D. Ferran, a renowned Vietnam sniper, who provided me with an illuminating account of his combat career. Jim O'Hern was also helpful in suggesting leads.

I received valuable assistance from several magazine editors, and I must thank John Elliott of *Soldier Magazine*, Tim Ripley of *Combat and Survival*, James Marchington of *Combat and Militaria*, and Robert K. Brown (and Dwight P. Swift) of *Soldier of Fortune*.

Chapter One

The Long Rifle and the American Revolution

Until the American Revolution (1775–83) the rifle had been used primarily as a hunting rather than a military weapon. Although the rifle was far more accurate than the smoothbore musket, it was slow to load and was generally considered too complex for the average, poorly trained infantryman to use effectively. But the order of the US Continental Congress on 14 June 1775 to raise ten companies of riflemen marked the first step in the development of the sniper, a rifle-armed soldier who used his skill as a marksman to kill selected individuals at long range.

The Congress was able to draw upon a fine tradition of rifle shooting. The long rifle used in the colonies had evolved in the mountains and backwoods of Pennsylvania, and although it owed its origins to the hunting rifles of central Europe, by 1775 the long rifle had become a separate and distinct weapon in its own right. The Kentucky rifle – as it subsequently became known – had a longer barrel and a smaller bore than the European jaeger or hunting rifle. As a result, it was more accurate and used less powder. An original Kentucky rifle underwent a series of trials in 1922, and placed five shots into a 2.1-inch group when fired from a rest at a range of 100 yards. Such figures were undoubtedly impressive by the standards of the late eighteenth century.

Marksmanship was highly prized by the American settlers, not only for obvious hunting reasons but also because it was a means for a man to display his prowess to his fellows. Competitions were commonplace, as Harold L. Peterson explains:

The German and Swiss colonists brought with them their love of shooting matches, and all along the frontier such contests flourished as one of the chief forms of recreation. There were turkey shoots with a live bird tethered behind a log as the target. The trick was to induce the turkey to show its head and then hit it. The first man to do so won the bird as a prize. There were also beef shoots. Here each contestant brought his own target, a board or shingle about five by seven inches with an X marked in the centre. The man to score the closest shot to the centre of his mark won the choicest cut of beef, and so on for each round until the carcass was completely disposed of. The ranges varied. Much depended upon the size of the target. A small one, such as the head of the turkey, might call for distances of two hundred and fifty feet offhand, or three hundred and fifty feet with a rest.[1]

As the colonial riflemen prized individualism over conformity, doubts were expressed among senior American officers about their suitability as soldiers. To the town dwellers of the Eastern seaboard, they presented a strange and exotic sight. Dressed in fringed hunting shirts and carrying long knives or tomahawks, they delighted in demonstrating their skills as marksmen to any who would watch. In one instance, involving two brothers, one of them held a piece of board between his knees with a paper mark the size of a dollar fixed to the centre, while the other fired eight consecutive shots through the board at a range of sixty yards.

Initially wary of employing frontiersmen as soldiers, George Washington, the Congressional Commander-in-Chief, eventually changed his mind. He wrote to Colonel Daniel Morgan, one of the rifle pioneers, to encourage their recruitment, while suggesting his own ideas on tactics:

The corps of rangers nearly formed, and under your command, are to be considered as a body of light infantry, and are to act as such. It occurs to me that, if you were to dress a company or two of true woodsmen in the Indian style, and let them make the attack with screaming and yelling, as the Indians do, it would have very good consequences, especially if as little as possible were said or known of the matter beforehand.[2]

Charles Lee, another senior officer in the Continental Army, also encouraged the formation of sharpshooting units. He wrote:

The frontier riflemen will make fine soldiers. [He praised] . . . their amazing hardihood, their methods of living so long in the woods without carrying provisions with them, the exceeding quickness with which they can march to distant parts, and, above all, the dexterity to which they have arrived in the use of the rifle gun. There is not one of these men who wish a distance less than 200 yards or greater object than an orange. Every shot is fatal.[3]

In the first phase of the war, when British forces were cooped up in Boston, riflemen quickly proved their worth, according to this passage from Charles Winthrop Sawyer:

In the army around Boston the riflemen were employed as sharpshooters to pick off any British soldiers or officers who were incautious in exposing themselves. This they did to perfection. There is mention of a British soldier shot at 250 yards when only half his head was visible; of ten men, three of whom were officers, killed one day while reconnoitring; of a rifleman who, seeing some British on a scow at a distance of fully half a mile, found a good resting place on a hill and bombarded them until he potted the lot.[4]

The accuracy of such accounts must be leavened with a degree of conjecture; in the course of retelling a shooting exploit there is a natural tendency for ranges to increase, and likewise the number of casualties. And yet Lord Howe, the British commander in

Boston, was sufficiently impressed by the American riflemen to complain to his government of the 'terrible guns of the rebels'.

The limited and costly British victory at Bunker Hill in 1775 was followed by generally indecisive manoeuvring; the British attempted to split the colonies in half but had little success. In 1777 an offensive was launched by the British to settle the war: General John Burgoyne led an army south from Canada to isolate New England from the other colonies. After some initial success, Burgoyne was soundly defeated in a series of engagements that culminated in the Battle of Saratoga.

The densely wooded and broken terrain of the Saratoga battlefield greatly helped the American riflemen, ably led by Daniel Morgan. Saratoga was one of the turning points of the war, and Morgan's riflemen played a key role in the American victory. Throughout the war the Americans were fortunate in finding commanders whose ability and resolution could dramatically affect the outcome of an engagement, as this description of Morgan suggests:

> Morgan had won a great reputation for bravery and resource in the French and Indian Wars. He was of Welsh descent, a native of New Jersey, but a resident of Virginia; his stature and his tenacity of purpose were equally immense; he was uneducated, but he had a clear and strong intelligence. With his Virginians he rose from obscurity to international fame. Before the close of the war Frederick the Great spoke of him as 'the greatest leader of light infantry in the world'.[5]

As Burgoyne's men advanced southwards down the valley of the Hudson River they came into contact with the Americans under the command of Generals Gates and Arnold. Lieutenant William Digby of the Shropshire Regiment left this record of his encounter with Morgan's riflemen on 19 September 1777:

> At day break intelligence was received that Colonel Morgan, with the advance party of the enemy consisting of a corps of

rifle men, were strong about three miles from us. A little after 12 our advanced picquets came up with Colonel Morgan and engaged, but from the great superiority of fire received from him – his numbers being much greater – they were obliged to fall back, every officer being either killed or wounded except one, when the line came up to their support and obliged Morgan in his turn to retreat with loss.[6]

Although Digby was understandably keen to stress that Morgan's men apparently outnumbered his own, what is significant is the deployment of the American riflemen, concentrating on the elimination of officers, and how, once British reinforcements arrived, they retired in the manner of experienced skirmishers. British ineffectiveness in dealing with the American sharpshooters was one of the features of the battle.

An American account of Morgan's riflemen in action is provided by Captain E. Wakefield, who had been attached to Dearborn's Light Infantry, and who had a clear oversight of the engagement on 19 September:

I shall never forget the opening scene of the first day's conflict. The riflemen and light infantry were ordered to clear the woods of the Indians. [General] Arnold rode up and, with his sword pointing to the enemy emerging from the woods into an opening covered with stumps and fallen timber, said, 'Colonel Morgan, you and I have seen too many redskins to be deceived by that garb of paint and feathers; they are asses in lions' skins, Canadians and Tories; let your riflemen cure them of their borrowed plumes.'

And so they did; for in less than fifteen minutes the 'Wagon Boy', with his Virginia riflemen, sent the painted devils with a howl back to the British lines. Morgan was in his glory, catching the inspiration of Arnold, as he thrilled his men; when he hurled them against the enemy, he astonished the English and Germans [fighting for the British] with the deadly fire of his rifles.[7]

Despite the contribution of the American riflemen, the fighting on 19 September led to stalemate; the British held their ground but suffered heavy casualties. A three-week lull followed until battle was resumed on 7 October. The Americans were aware that of the British commanders opposing them, General Simon Fraser was the most able. Through the fog of musket smoke, Fraser was spotted by General Arnold. Samuel Woodruff, an American volunteer, relates the subsequent incident which changed the course of the battle:

> Soon after the commencement of the action, General Arnold, knowing the military character and efficiency of General Fraser, and observing his motions in leading and conducting the attack, said to Colonel Morgan, 'That officer upon a grey horse is of himself a host, and must be disposed of – direct the attention of some of the sharp-shooters among your riflemen to him.'
>
> Morgan nodded his assent to Arnold, repaired to his riflemen, and made known to them the hint given by Arnold. Immediately upon this, the crupper of the grey horse was cut by a rifle bullet, and within the next minute another passed through the horse's mane a little back of his ears.
>
> An aide of Fraser, noticing this, observed to him, 'Sir, it is evident that you are marked out for particular aim; would it not be prudent for you to retire from this place?'
>
> Fraser replied, 'My duty forbids me to fly from danger,' and immediately received a bullet through his body. A few grenadiers were detached to carry him to the Smith house.[8]

The rifleman who downed Fraser from his exposed position is thought to be Tim Murphy, a legendary Indian fighter and one of Morgan's best marksmen. The consequences of Murphy's shot were calamitous for the British, as recorded by Lieutenant Digby:

> Brigadier General Fraser was mortally wounded, which helped to turn the fate of the day. When General Burgoyne

saw him fall, he seemed then to feel in the highest degree our disagreeable situation. Our cannon were surrounded and taken – the men and horses being all killed – which gave them additional spirits. It evidently appeared a retreat was the only thing left to us.[9]

The defeat of Burgoyne not only thwarted the British attempt to separate the Northern colonies, but persuaded the French to declare war on Britain and send ships and troops to the new republic's aid.

Despite the inflexibility of much of the British military system, within the British Army there were officers with experience of conditions in North America who were determined to introduce light infantry tactics to counter the American frontiersmen. J.G. Simcoe had previously fought against the French in a unit of irregulars called the Queen's Rangers, and recommended the development of light troops to fight in the American interior. He made this request, written in the third person, on the need for suitable clothing for the purposes of camouflage:

It having been determined, for the next year, to clothe the Provincials in red, Major Simcoe exerted to preserve the Rangers in green, and to procure for them green waistcoats, [for] green is without comparison the best colour for light troops with dark accoutrements; and if put on in the spring, by autumn it nearly fades with the leaves, preserving its characteristic of being scarcely discernible at a distance.[10]

In August 1776, shortly before his death, General Fraser had issued a memorandum calling for the raising of a company of marksmen to accompany the main force during operations prior to the Saratoga campaign:

A man from each company of the British regiments to form a corps of marksmen consisting of a hundred men. They should be chosen for their strength, activity and being expert at firing ball; each man to be furnished with an excellent firelock, the lock in good order and the hammer well steeled. The soldier should by his frequent experience find out the quantity of powder with which his firelock fires the justest at the greatest distance and his cartridges should be made by that measure. Officers of experience should be appointed to this company.

Monin's (Canadian) Volunteers to be augmented to seventy or eighty and provided with clothing, arms, accoutrements. These two corps may act on the flanks of the advanced Brigade and reinforced by what number of Indians the General may think fit to employ. They may be turned to great use as the nature of the country can admit of their turning large corps of the enemy, surprising convoys and making them uneasy in their rear: a corps of this kind well commanded would fatigue the enemy exceedingly by constant alarms.[11]

Fraser's memorandum was acted upon with surprising speed and in September a company of sharpshooters was formed, commanded by Fraser's nephew, Captain Alexander Fraser. The unit did not last long, however, being virtually wiped out at the Battle of Bennington on 16 August 1777. It was one the many ironies of the war that a proponent of light infantry like General Fraser should be the victim of a marksman's bullet.

A significant problem faced by British light infantry reformers was the poor quality of the British musket, popularly known as the 'Brown Bess'. This account reveals its many shortcomings:

The Brown Bess [was] the very clumsiest and worst contrived firelock in the world. It required the largest charge of powder and the heaviest ball of any; yet owing to the absence of every scientific principle in its construction, its weight and windage were the greatest, its range the shortest, and its accuracy the least; at the same time that it was the most costly of any similar arm in use, either in France, Belgium, Prussia or Austria. Taking a long series of engagements, only one musket ball out of 460 was found to take effect.[12]

Colonel George Hanger, a British firearms expert and veteran of the American Revolution, also disparaged the Brown Bess, declaring that a good example might hit a man at 80 yards, but 'as to firing at a man at 200 yards, with a common musket, you may as well just fire at the moon'. Hanger had experience of being fired on by American riflemen, and he thought that a crack shot could hit the head of a man at 200 yards, and 'provided an American rifleman were to get a perfect aim at 300 yards at me, standing still, he most undoubtedly would hit me, unless it was a very windy day'.

Although Hanger had the regular soldier's traditional distaste for militia and irregulars, he accepted the reason for the frequent setbacks suffered by the British at the hands of the colonials being that 'they were better marksmen than our regulars'. The success of American arms owed much to, 'the unerring and surprising skill of the American backwoodsman, who, from the age of sixteen, has made the use, perfection, and construction of the rifle, and all other species of arms, both his study and his pleasure'.[13]

The British made repeated, though generally fruitless, attempts to acquire rifles and men trained in their use. As a stopgap measure, rifle-armed jaegers were imported from the German states, but their weapons compared poorly with the Kentucky long rifle, and the general quality of German marksmanship was poor.

A British officer, Captain Patrick Ferguson, produced his own breech-loading rifle, which for the times was a very advanced design. During trials in 1776, Ferguson impressed the War Office with the accuracy and rate of fire of the rifle, and an experimental corps was despatched to America. In the field, however, the rifle and its inventor proved less effective. Ferguson was badly wounded at Brandywine Creek and after a slow recovery he returned to service, only to be killed by an American rifleman at the battle of King's Mountain in 1780. His corps was disbanded and the rifle withdrawn.

A last word on the success of the American rifleman was provided by Captain Henry Beaufoy, attempting to encourage the introduction of rifle-armed troops into the British Army during the Napoleonic Wars:

> The Americans, during their war with this country, were in the habit of forming themselves into small bands of ten or twelve, who, accustomed to shooting in hunting parties, went out in a sort of predatory warfare, each carrying his ammunition and provisions, and returning when they were exhausted. From the incessant attacks of these bodies, their opponents could never be prepared, as the first knowledge of a patrol in the neighbourhood was generally given by a volley of well-directed fire, that perhaps killed or wounded the greater part.[14]

Chapter Two

Sharpshooting in Europe

The British Army introduced light troops into the infantry during the course of the eighteenth century. But the war against the colonists in America forced the British to consider the possibility of raising rifle-armed troops to act as marksmen. While some authorities still regarded riflemen as gamekeepers or hunters, with no place on the battlefield, the many reverses suffered in America strengthened the hand of the military reformers and gave them the chance to experiment with rifle units.

Captain Henry Beaufoy published a major treatise on the nature and use of rifles in war, which included a commentary on the psychological effects of aimed fire. Beaufoy wrote of the fear induced in soldiers by enemy riflemen:

> It has readily been confessed to the writer by old soldiers, that when they understood they were opposed by riflemen, they felt a degree of terror never inspired by general action, from the idea that a rifleman always singled out an individual, who was almost certain of being killed or wounded; and this individual every man with ordinary self-love expected to be himself. Destroy the mind, and bodily strength will avail but little in that courage required in the field of battle.[1]

Beaufoy's remarks could as readily be applied to the sniper of the two world wars as the sharpshooter of the late eighteenth century;

undermining enemy morale has remained one of the constants in the history of sniping.

The outbreak of war with France in 1793 (which lasted until 1815) came at a time when the British Army was still developing its ideas on rifle troops, and had yet to put them into practice. The defeats suffered by British in the the 1790s — repeatedly outmanoeuvred by sharpshooting French skirmishers — only served to underline the requirement for trained marksmen.

The failings of the British were to a large degree a consequence of nationalist complacency; officers considered their light companies to be the match of their French opponents, but failed to realize that the French had revitalized and transformed infantry tactics. In many instances, British light infantry were so in name only. These weaknesses were documented by General Money, who wrote:

> British Light Infantry are not, in every instance, equal to the service performed by Light Troops. Irregulars, in most services in Europe, are furnished with a rifle with which they are taught to fire with great exactness. Will any British Officer, who knows the use of firearms, say, that, with one of the muskets used by our Light Infantry, he could fire as accurately at a mark, as with a park-keeper's rifle? Chasseurs in foreign service are taught never to waste a shot, and whenever they do fire, it rarely happens but a man is killed or wounded: they are taught to conceal themselves as much as possible; to creep from bush to bush, and if pressed to run off, for retrograde motions are not deemed disgraceful to Chasseurs.
>
> A [British] Light Infantry Man fires where he sees smoke, and continues firing till he has wasted all his ammunition. This is nine times out of ten the case. All this proceeds from his not being trained as a marksman. When a Light Infantry Man hears a ball pass him, which he has had the good luck of having escaped, he turns and fires at random; instead of concealing himself as a rifleman would do, and looking for the man who fired at him.[2]

Money's criticisms of British practice were echoed by William Surtees. A British soldier of long experience who eventually served in the famous 95th Rifles, Surtees made this plea for marksmanship in the light of adverse comment from the New World:

> The Americans tauntingly tell us, our soldiers do not know how to use the weapons that we put into their hands; and truly, if we are to judge by the awful destruction which they have occasionally inflicted upon our brave soldiers, we should be led to expect that they understand this science much better than ourselves.
>
> It might, however, be easily remedied, if more attention were paid to the instruction of the recruit in this most essential qualification, and more time and ammunition devoted to target practice; but at the same time every officer should be made to know something of the projectile in general, or he will be incapable of instructing his men. I will venture to assert, that eight out of ten of the soldiers of our regular regiments will aim in the same manner at an object at the distance of 300 yards, as at one 50. It must hence be evident that the greater part of those shots are lost or expended in vain; indeed the calculation has been made that only one shot out of two hundred fired from muskets in the field takes effect, while one out of twenty from rifles is the average.[3]

During the late eighteenth century there remained the belief that the British were unsuited to skirmishing and marksmanship, skills that came more naturally to Americans or Continental Europeans. One reason for this was advanced by Colonel George Hanger, who explained:

> Now the German Jäger, brought up in the forests to shooting at everything with the rifle since the age of fifteen, is taught all distances by the practice of years: for he can never offer himself to service as a Jäger, unless he can produce certificates

from the masters of the forests that he has served an apprenticeship of seven years, and is a perfect shot. A British soldier can never be taught to be a perfect judge of different distances. Jägers from their natural servitude, and great practice, are in no want of being taught distances; the knowledge and precision of judging different distances comes to them naturally from practice, from their early youth.[4]

In August 1799, a British force landed in northern Holland, and in the ensuing campaign the faults of the British were ruthlessly exposed by French skirmishers. During the landings and subsequent fighting at Helder the British suffered heavily. The misfortune was recorded in the diary of Sir Henry Bunbury:

The enemy from the crests of the sand-dunes kept up a constant and destructive fire, while he was himself sheltered by their folds from the guns of the British. The loss of the enemy in killed or wounded was probably small: but the disadvantages under which the invaders fought necessarily exposed the three brigades engaged to a serious loss of men; and a large proportion of the superior officers of the staff fell under the aim of the riflemen. Two Lieut-Colonels and about fifty men were killed on the spot: and Lieut-General Sir James Pultney, five field officers and nearly 400 others were wounded.[5]

The effectiveness of aimed fire at selected, high-value targets was also recorded by Henry Beaufoy; officers were forced to disguise their badges or distinctions of rank (a precautionary measure also adopted in later wars): 'When opposed to riflemen, it is the bravest who fall, for it is the bravest who expose themselves most, and thus become the most conspicuous. The officers of our Army in Holland obtained this experience, and in several instances found it necessary to change their hats, and assimilate themselves to the private men.'[6]

At the Battle of Bergen in 1799, British forces in Holland

again found themselves at a disadvantage when faced by the French. William Surtees, then a soldier in the 56th Regiment of Foot, provided this account of the battle:

> After the fighting had fairly commenced, we kept but little order, owing partly to the want of discipline and experience in our people, and partly to the nature of the ground, which was rugged and uneven in the extreme, being one continual range of sand-hills, with hollows, more or less deep between them; and partly it may be attributed to the ardour of our young men, who pressed on perhaps too rapidly.
>
> We continued to advance, and never once made a retrograde movement, the enemy regularly retiring from height to height on our approach; but they had greatly the advantage over us in point of shooting, their balls doing much more execution than ours; indeed it cannot be wondered at, for they were all riflemen, trained to fire with precision, and armed with a weapon which seldom fails its object if truly pointed; while we were (what shall I say) totally ignorant of that most essential part of a soldier's duty.
>
> They consequently suffered little from our fire; but we could not believe this, and tried to persuade ourselves that they had either buried their dead in the sand before we came up to them, or carried them off as they retreated; but experience has since taught me to know that we then must have done them little harm.[7]

The usefulness of riflemen in offensive operations was recorded by Captain Beaufoy. During the 1799 Russian offensive against the French Army in Switzerland, riflemen proved invaluable to the French cause:

> Near the famous pass of the Pont du Diable, the Russians had erected a battery, rendering its passage almost impracticable, and the French were reduced to the greatest dilemma possible, till a party of Swiss riflemen, serving as volunteers

in the army, offered their services, which were readily, but not without anxiety, accepted. These immediately posted themselves with the utmost deliberation on the top of a rock, at a distance of about 400 yards, which completely laid open the cannoneers, who were working the guns in the battery below. It is needless to say, that in the course of a very short time, the whole of them were either killed or wounded, and the French lost not a man in doing that, which, had they had recourse to the ordinary method, would have been attended with great difficulty and loss of blood and time.[8]

Despite the many setbacks experienced by the British, during the early 1800s an effective corps of riflemen was raised. The first rifle-armed unit was the fifth battalion of the 60th Foot, jaegers raised from the many Germans in British military service. This was followed by the 95th Rifles, eventually consisting of three battalions, who fought with great distinction in the Peninsular War. Fortunately for the British, the commanders of these troops included General Sir John Moore and Colonels Coote-Manningham and Stewart, men of enlightened approach who proved excellent trainers of light infantry.

In 1800, trials were held to find a suitable rifle for the Army. The winner was a rifle submitted by Ezekiel Baker, a gunsmith from Whitechapel. The Baker rifle – with a calibre between .615 and .70 of an inch – was not particularly accurate and used large quantities of gunpowder, but it was reliable and sturdy, and for those reasons it was chosen by the Ordnance board. The short 30-inch barrel was augmented by a 24-inch sword bayonet, thus giving a rifleman parity of length with an opponent armed with a conventional musket and bayonet.

The previously almost unknown art of marksmanship was emphasized in training, as revealed in this entry from the order book of the 5/60th:

The true rifleman will never fire without being sure of his man. And he will recollect that a few well-directed shots, that tell, will occasion greater confusion than thousands fired at random and without effect, which will only make the enemy despise our fire, and inspire him with confidence in proportion as he finds us deficient in skill and enterprise.[9]

Captain Beaufoy cautioned his readers that rifles must be used at ranges beyond that of the smoothbore musket:

A rifle should begin where a musket ceases to be of use; and unless riflemen are kept at least 250 yards from a line of musketry, the latter will by their greater facility and expedition in loading and firing drive the former out of the field; the number of balls from a musket being at least double or treble to those fired in equal spaces of time from a rifle.[10]

The 95th Rifles held marksmanship in high regard, as William Surtees discovered when subjected to a test of his shooting abilities as a new recruit in the regiment. Under the eye of his instructor, Major Wade (subsequently commander of the 2nd Battalion), Surtees was instructed to fire at the target:

I did and made a pretty good shot, hitting near the bull's eye, on which he made me load again and fire, and hitting that also, he made me go on till I had fired ten rounds, all of which hit the target, and two of which had struck the bull's eye. The distance indeed was only fifty yards, but for a recruit, unaccustomed to rifle shooting, he called it a wonderful exhibition, and in consequence he gave me six-pence out of his pocket, and ordered me home.

The officer (Major Wade) was one of the best shots that I have almost ever seen. I have known him, and a soldier called Smeaton, hold the target for each other at a distance of 150 yards, while the other fired at it, so steady and accurate was their shooting.

I continued to maintain my character as a good shot,

which in a rifle corps is a great recommendation, and proceeded in acquiring a knowledge of my duties with such steadiness as obtained the approbation of my officers. I was employed in doing the duty of one having been appointed what is termed a 'chosen man'.[11]

Fighting against the French in the Peninsular campaigns, the British riflemen gained great renown. Dressed in dark green uniforms with black facings, the 95th fought in almost all the major engagements in Portugal and Spain. The bulk of the British Army continued to experience difficulties when fighting in a skirmishing role, but the 95th regularly outfought French voltigeurs and chasseurs. Lieutenant Blakiston, a light infantry officer of the 43rd Foot, generously accorded the 95th high praise: 'They possessed an individual boldness, a mutual understanding of the ground I never saw equalled. They were in fact superior to the French voltigeurs as the latter to our skirmishers in general.' The weakness in the British infantry was also noted by Edward Costello, a rifleman in the 95th. During the battle of Fuentes de Onoro he recorded how one British battalion had been badly mauled by the French:

> Our next attempt was upon the left of Fuentes, where one company was detached, while the remainder of the regiment was ordered to take possession of the town. The section to which I belonged were posted near the banks of the River Dos Casa. The 70th Highlanders had suffered very severely here, as the place was strewn about with their bodies. Poor fellows! they had not been used to skirmishing, and instead of occupying the houses in the neighbourhood, and firing from the windows, they had, as I heard, exposed themselves, by firing in sections.[12]

At the siege of Badajoz in 1812, the British encountered exceptionally strong French fortifications. Costello records how the

95th were used as marksmen to eliminate key targets on the French walls:

> A very disagreeable duty, that usually fell upon a few of the best shots of the battalion, consisted in being obliged to run out, in independent files, to occupy a number of holes, that had been dug at night between our batteries and the walls of the town. From these pits, of which each man had one to himself, our particular business was to pick off any enemy who exposed themselves at their guns, on the walls through the embrasures. Many a Frenchman was thus knocked off by us. But it often occurred also that our men were killed or wounded in their holes, which made it doubly dangerous for the man of the relieving party, who instead of finding a ready covering, perceived it occupied by a wounded or dead man. Before he could get shelter therefore or remove the body, there was a great chance of his being shot.[13]

Other writers commented on the use of marksmen to shoot enemy artillerymen. George Simmons wrote of how a unit of riflemen – 'as prime fellows as ever pulled a trigger' – silenced a French battery, and Lieutenant-Colonel J. Leach also recorded their success:

> Some riflemen were ordered to approach as near as possible to the Guadiana and to endeavour, by shooting the artillery-men across the river, to silence some field pieces brought out by the French from Fort San Christobal, which enfiladed the [British] trenches. This had the desired effect; and the field pieces were withdrawn into the fort, after some of the gunners had bitten the dust.[14]

The confusing and dangerous nature of the Napoleonic battlefield, where a cannon-ball could take off a man's head in an instant and where the troops were virtually blinded by clouds of gun smoke, is vividly recreated by John Harris, a rifleman in the 95th. He describes his part in the battle of Vimeiro:

> I was myself very soon so hotly engaged, loading and firing away, enveloped in the smoke I created, and the cloud which hung about me from the continued fire from my comrades, that I could see nothing for a few minutes but the red flash of my own piece amongst the white vapour clinging to my very clothes. The Rifles, as usual, were very busy in this battle. The French, in great numbers, came steadily down upon us, and we pelted away upon them like a shower of leaden hail. Under any cover we could find we lay; firing one moment, jumping up and running for it the next; and, when we could see before us, we observed the cannon-balls making a lane through the enemy's columns as they advanced, huzzaing and shouting like madmen.[15]

Conditions such as these made marksmanship almost impossible; individual skirmishing was conducted during the opening stages of the battle or in areas away from the main fighting. Harris did not consider himself a crack shot — he was a cobbler in the battalion — but when the fighting at Vimeiro had almost ceased he found an opportunity to use his rifle for aimed fire. Looting was a common practice among all armies of the period, and at the close of battle, soldiers would roam the field to plunder the dead and dying. This was a recognized means of replenishing supplies and augmenting the private soldier's meagre income. Harris had tried on a pair of shoes belonging to a dead British officer in the 50th Foot:

> As I did so I was startled by the sharp report of a firelock, and, at the same moment, a bullet whistled close by my head. Instantly starting up, I turned, and looked in the direction from whence the shot had come. There was no person near me in this part of the field. The dead and the dying lay thickly all around; but nothing else could I see. I

looked to the priming of my rifle, and again turned to the dead officer of the 50th. It was evident that some plundering scoundrel had taken a shot at me, and the fact of his doing so proclaimed him an enemy. To distinguish him amongst the bodies strewn about was impossible; perhaps he himself might be one of the wounded.

Hardly had I effected the exchange, and put on the dead officer's shoes, and resumed my rifle, when another shot took place, and a second ball whistled past me. This time I was ready, and turning quickly, I saw my man: he was about to squat down behind a small mound. I took a haphazard shot at him, and instantly knocked him over. I immediately ran up to him; he had fallen on his face, and I heaved him over on his back, bestrode his body, and drew my sword bayonet. There was, however, no occasion for the precaution as he was even then in the agonies of death.

It was a relief to me to find I had not been mistaken. He was a French light infantryman, and I therefore took it quite in the way of business – he had attempted my life, and lost his own. It was the fortune of war; so, stooping down, with my sword I cut the green string that sustained his calibash [water bottle] and took a hearty pull to quench my thirst.[16]

The last campaign involving the 95th Rifles during the Napoleonic Wars culminated in the battle of Waterloo in 1815. The 95th maintained the defence of the key position of La Haye Sainte for most of the battle, making the accuracy of their fire tell against the advancing French. One of the riflemen present at Waterloo, Private Wheeler, wrote this letter describing an incident in the battle:

I was ordered with two men to post ourselves behind a rock or large stone, well studded with brambles. This was somewhat to our right and in advance. About an hour after we were posted we saw an officer of Huzzars sneaking down to get a peep at our position. One of my men was what we

term a dead shot, when he was within point-blank distance.
I asked him if he could make sure of him. His reply was 'To
be sure I can, but let him come nearer if he will, at all events
his death warrant is signed and in my hands if he should
turn back.' By this time he had without perceiving us come
up near to us. When Chipping fired, down he fell and in a
minute we had his body with the horse in our possession
behind the rock.[17]

In the long peace that followed the end of the Napoleonic Wars
improvements in firearms design made rifles more reliable and
quicker to reload. The development of the expanding Minié bullet
and the percussion lock led to the demise of the old smoothbore
flintlock musket and its replacement – from the 1840s onwards –
by the rifle. In 1853, the British Army adopted the Enfield, a
muzzle-loading rifle which was sighted to a distance of 1200
yards, although ranges half that distance would have been beyond
all but a crack shot. Despite the possibilities of long-range
marksmanship offered by rifles such as the Enfield, European
armies paid little attention to improving standards in accurate
shooting.

Limited numbers of Enfields were issued during the Crimean
War (1853–56), but only in the Indian Mutiny of 1857 were
they used in quantity. Rumours that the grease used to ease the
Enfield bullet down the barrel came from animal fat outraged
Hindu and Muslim troops serving the British, and was a contrib-
utory cause in the mutiny against British rule. Large areas of
northern India fell to the mutineers, including the city of
Lucknow. Only the Residency in Lucknow remained in British
hands, and was subjected to a prolonged siege. This account
details the activities of sharpshooters during the defence of
Lucknow:

> Linked together by hastily contrived defences, the stronghold
> was manned by a garrison of about 1700, of whom not more
> than one half were Europeans. The only way fire could be

returned was from concealment. The garrison came gradually to encourage the assailants to occupy a point and to have confidence in occupying it. But they marked well the direction; and during the night they bored holes in that direction. In the morning the enemy would come up by twos and threes to occupy their chosen post. Then the muskets would be discharged. The result was almost inevitably successful.

But there was a wily marksman whom this guileful device entirely failed to snare. Choosing his coign of vantage with infinite subtlety and care, he would await the appearance of a likely target with an unblinking watchfulness that dawn-chill or the heat of the noonday sun in no way seemed to affect. Unseen, undetectable, he would remain inactive for hours at a stretch. But if a gleam of scarlet tunic or white cap cover came to reward his vigil, then his swift, unerring shot took toll with a deadly precision that scorned the waste of a single cartridge.

In wry tribute to his outstanding skill, the sweating men of the garrison dubbed him 'Jim the Nailer'. To give the exact tally of his score against the Residency's garrison is as impossible as to cloak him with a specific identity or pronounce upon his ultimate fate. For with the relief of Lucknow, no more was heard of him.[18]

The activities of 'Jim the Nailer' had no effect on the British tendency to concentrate on training soldiers in disciplined volley fire at the expense of individual marksmanship. During one of the campaigns against the Pathan tribesmen on the North-west Frontier of India, the British suffered heavy casualties in an engagement that demonstrated the shooting ability of the Pathans armed with single-shot Martini rifles. Of note is the manner in which each Pathan marksman was served by several loaders to increase his rate of fire. Written by General Sir George MacMunn, then a gunner subaltern, this narrative describes how a brigade of British infantry was ordered to assault a ridge held by the Pathans:

The tribal standards waved on the troops and defiant cries
were wafted down and re-echoed in the gorge. Colonel
Travers led his first few scouts across easily enough to some
cover under overhanging rocks. Then the defenders awoke to
what was in progress. The remainder of the Gurkhas,
attempting to join their colonel in extended order along the
narrow neck, encountered a tremendous fire, chiefly of
Martini bullets aimed by the best marksmen on the frontier.
Every expert in the clefts above had two or three loaders.
Hardly a shot missed its billet. The men, dribbling over,
were hit time and again, and rolled down the slopes on either
side or lay on the fairway. Colonel Travers had hoped, as
soon as a fair clump of men joined him, to begin pushing up
the goat tracks where there appeared to be occasional cover
from rocks. But his party did not increase. The accurately
aimed fire swept all who ventured to join him.[19]

The failure of the Gurkhas to gain the ridge led the brigade
commander to send his other battalions into the attack, first the
Dorsets and then the Derbyshire Regiment. Again, the assaults
were halted by Pathan fire. A final push led by the Gordon
Highlanders had more success, and as the British neared the
summit, 'the heavy rifle fire soon died away, and the leading files
gained the top at various parts almost unmolested. The tribesmen
had seized their standards and had gone.' The Gordon Highlanders
alone had suffered 300 casualties, which bore testimony to the
accuracy of the Pathan's marksmanship.

The Franco-Prussian War (1870–71) was characterized by
battles of manoeuvre where mass firepower outweighed individual
marksmanship, but once the Prussian Army began to besiege
Paris then sharpshooting become increasingly important. Armed
with breech-loading Dreyse Needle guns, the Prussians engaged
in sniping duels with the French, who were equipped with the
superior Chassepot breech-loader. While breech-loading rifles
increased the overall rate of fire, their prime advantage for snipers
was their ease of reloading. The old muzzle-loader was cumber-

some and noisy, and thus could give away a sniper's position; operating a breech-loader necessitated minimal movement and produced little noise.

The following account from the Franco-Prussian War emphasizes the skill and patience required by a marksman when faced by a wily opponent who was also a crack shot:

The area between Rainey and Ville Evrart, on the eastern outskirts of the capital, was held by a regiment forming part of Montbe's 12th Corps; and it was their unhappy fortune to be particularly victimized. From the upper room in the gardener's cottage of the Château Launay, a French sniper had bowled over man after man, shooting from such a range that effective return fire from the Germans was out of the question. Completely baffled how best to deal with the situation, the major commanding the battalion was only too willing to give his consent when a young cavalry officer, Ensign Baron Steinfurst-Wallenstein, offered to try and stalk the enemy marksman and catch him at a disadvantage. An expert gameshot, he had brought his favourite rifle with him on service and he was convinced that his trusty weapon could be relied upon to even up the score.

Passing through the snow-bound outposts before it was light, the Baron took up position in a clump of evergreens at an oblique angle to the gardener's cottage, whose glassless upstairs window gaped black and empty looking. With the wintry sun slowly clearing the haze, the watcher could just perceive a hint of movement at the back of the room, from whose shadows the sniper obviously kept observation on the German outpost line curving about Rainey's outskirts.

Three times during the day the flash of a shot lit the bedroom's December darkness, and three times the watcher glanced eagerly along his sights, hoping for a glimpse of the man who kept himself so carefully hidden. Once, there was the momentary gleam of a gun-barrel and the ensign's finger almost tightened about the trigger. But realizing the unlikelihood of ever getting a second shot should the first fail to find its target, the Baron soberly held his fire; and dusk found him stealing back to his own lines, far from discouraged and fully determined to resume his vigil on the morrow.

There was a good deal of rather bitter chaff when Steinfurst-Wallenstein reported back at battalion head-quarters; for one dead and two wounded men bore painful tribute to the sniper's unimpeded activity. But the Baron took it all without comment; and dawn found him again at his post of observation.

On the second evening, when Ensign Baron Steinfurst-Wallenstein reported back, the battalion commander curtly ordered him to return to his regular duties. 'Zu befehl, Herr Major,' the young officer dutifully resounded, 'but when it is light enough for you to use your field glasses, I think you will find that my time has not been wasted.'

Still slightly incredulous, the battalion commander was standing by at first light to focus his binoculars on the gaping window of the gardener's cottage. From it hung down the body of a man, pitched head foremost over the sill; his rifle released from his nerveless hands, making a long, dark streak against the snow-drift piled up against the wall below.

As Steinfurst-Wallenstein subsequently explained, the sniper had fired three times during the previous day, loosing off his final cartridge just before dusk. As always, he had aimed and fired from the back of the room; but in the fading light he had risked coming forward to see if his last shot had taken effect; thrusting his head and shoulders through the blank window space to peer across at the German outposts. It was at this moment that the Baron had brought his sights to bear; the opportunity for which he had waited so long and patiently had come at last; and the account was closed with a single pull of the trigger.[20]

Chapter Three

The American Civil War

The American Civil War (1861–65) was a golden age for the military rifle. Although relatively slow to load and fire, improvements in weapons technology had made the rifle a highly accurate weapon. The Crimean War (1853–55) hinted at the tactical importance of the rifle, but the American Civil War demonstrated its primacy on the battlefield.

Rifles could now deliver accurate fire at ranges in excess of four hundred yards, which meant that the frontal assaults of the horse and musket era were doomed to disaster. This simple yet profound transformation in the conduct of war was not realized at the time, and massed columns of brightly uniformed soldiers were regularly slaughtered by devastating fire from a single line of riflemen. In time, machine-guns and quick-firing long-range artillery would wrest away that primacy, but in the middle half of the nineteenth century the rifle reigned supreme.

The advances in rifle design were matched by the skills of the men firing these weapons. The sharpshooting abilities found in America were based on a long hunting tradition, which had proved its military worth during the American Revolution and the War of 1812. In other nations, recruits had to be extensively trained in the rudiments of shooting. In America, however, there was a large pool of expert riflemen readily available to the armies of North and South, and these men were swiftly co-opted as sharpshooters.

Following the outbreak of hostilities in the spring of 1861, units of all types were raised by prominent and patriotic individuals. In the North, Hiram Berdan, the target-shooting enthusiast and firearms expert, began recruiting a corps of riflemen to act as elite skirmishers for the armies of the Union. A dedicated self-publicist, Berdan was tireless in promoting his cause, and eventually he was given permission to form two regiments, the 1st and 2nd Regiments of US Sharpshooters.

Berdan looked to find suitably qualified men from all the states of the Union. Local papers were employed to drum up support for the recruitment campaign; they described the qualities required of a sharpshooter and speculated on the likely duties of a successful candidate. The *New York Post* of 4 June 1861 published this appeal to arms:

> Mr H. Berdan of this city calls for volunteers to form a corps of skirmishers for the Army, which shall consist of the best marksmen in the country. Such men as Berdan proposes to enlist in the corps of skirmishers will be able to kill a man at the distance of a quarter of a mile.
>
> The Corps of Sharpshooters will be used not in the midst of battle, but on the outskirts, where, beyond the smoke and fury of the engagement, they will be able to act independently, choose their objects, and make every shot tell. Posted in small squads at from one-eighth to three-eighths of a mile in the field, firing a shot a minute, and hitting their mark with almost dead certainty, they will be a great annoyance to the enemy. They will combine their attention to the officers and by picking these off, will bring confusion to the enemy's line.[1]

Prominent individuals in the various states were deputized by Berdan to raise men for the Sharpshooters. William Ripley from Vermont was responsible for the selection and training of a company of Sharpshooters, and subsequently wrote of his command during the war. Berdan and his fellow amateur soldiers drew up stringent rules for selection, as Ripley explained:

It was required that a recruit possess good moral character, sound physical development, and in other respects come within the usual requirements of army regulations. But as the men were designed for especial service, it was required of them that before enlistment, they justify their claim to be called Sharpshooters, by a public exhibition of their skill as would fairly entitle them to the name, and would warrant a reasonable expectation of usefulness in the field.

To ensure this, it was ordered that no recruit be enlisted who could not, in a public trial, make a string of ten shots at a distance of two hundred yards, the aggregate measurement of which should not exceed fifty inches. In other words, it was required that the recruit be able to place ten bullets in succession, within a ten-inch ring at a distance of two hundred yards.

Any style of rifle was allowed, however use of telescopic sights was not permitted. The applicant was allowed to shoot from any position, as long as he shot [offhand] from the shoulder.[2]

During the summer of 1861, selection trials were held throughout the Northern states. The opportunity of joining a devil-may-care unit of elite infantry ensured a large response from young men who fancied their chances as crack shots. Although, as one veteran sharpshooter warned, there was a difference between firing on the range and the real thing: 'Shooting skill may fail even a champion marksman in battle when bullets pass right over or next to [one's] head [and] when [studied] calculation of distance is out of the question because positions are changed from minute to minute.'[3]

Despite Berdan's rigorous criteria for selection, some of the recruiting officers adopted a more easygoing approach. One captain in Minnesota was prepared to accept men who had little knowledge of firearms: 'Let no good man stand behind because he is not used to the rifle.' In the main, however, the men who joined the two regiments of US Sharpshooters were not only highly motivated but excellent marksmen. Once enlisted, they needed to be armed and equipped, and turned into soldiers.

During the first year of the war, the Sharpshooters remained out of action, and to the disgust of the more hotheaded they spent most of their time in camp. Charles Stevens was a Sharpshooter who later became the Regiment's historian. He drafted this description of life in camp:

About nine o'clock we had guard mounting. Drills, company or regimental, occurred twice a day, and therein the Sharp-shooters made a fine appearance. We had a proficient instructor, and apt and careful company officers. Among them Captain Drew, who was thoroughly conversant with tactics before entering service. He was popular throughout the entire 1st Regiment.

Towards the close of the afternoon, before retreat, and weather permitting, dress parade was held. This was a popular feature of camp life, witnessed generally by many spectators, and really a grand performance. Here reports were made as to the condition of the companies, whether they were all present and accounted for. Orders were read by the adjutant, and inspections and reviews frequently occurred.

Sunday morning inspections included dress, general appearance, packed knapsacks, etc. During the forenoon, church call brought the entire regiment, excepting those on duty, to the parade ground, where the Chaplain officiated. In bad weather these duties and services were dispensed with.

Besides target shooting other diversions were indulged, in order to train the men in the arduous duty of active service, particularly marching, destined to be often long and fatiguing. Therefore, football, jumping, racing, wrestling, boxing, fencing were prominent, which seemed to keep up a good feeling among the men, and between the companies.[4]

The mass of recruits included men who had recently emigrated from Central Europe, where target shooting and hunting with the rifle had a long pedigree. The Swiss were particularly well represented. One of their number was Rudolf Aschmann, who joined A Company, 1st US Sharpshooters on 20 August 1861. A keen recruit, Aschmann was pleased with the distinctive uniform issued to his unit:

> We had simple green coats which were quite wearable even as civilian clothes, green and blue pants of woollen cloth, a wide-brimmed felt hat as protection against the rain as well as the scorching rays of the sun, broad-soled, comfortable shoes, flannel shirts, and woollen socks. Everything, although not fitting snugly, was yet becoming, and a man was not constrained by either necktie or straitjacket.
>
> The officers' uniforms, though made with more care and of finer cloth, were not much different from those of the soldiers and just as simple by comparison. The insignia were no glittering epaulettes, only a narrow band edged with gold braid and fastened on the shoulder, showing the different badges of rank. For daily use or so-called lesser tenue we had a blue flannel jacket which was worn in the field even by officers.[5]

The choice of a suitable firearm for the Sharpshooters was a problem for Berdan. Recruits were encouraged by a $60 bounty to provide their own target rifles, which, although very accurate, were too heavy (some weighed as much as 32 lb) and too delicate to survive as a service rifle. One Sharpshooter wrote:

> It was soon found that there were severe objections to the use of these heavy weapons in the field, [which had been] so effective on the target ground. The great weight of them was almost prohibitory, for to a soldier burdened with the weight of his knapsack, haversack, canteen, blanket and overcoat, the additional weight of a target rifle was too much to be easily borne.[6]

Berdan vacillated over the correct choice of weapon. The US Army authorities wanted him to accept the infantry's standard rifle, the Springfield, which was a very basic rifled muzzle-loader. For a while, Berdan accepted this choice, but subsequently changed his mind to request the far superior M1859 Sharps breech-loader. Not only more accurate than the Springfield, the Sharps had a swifter rate of fire, and thus was eminently suitable for Sharpshooters acting in a skirmishing role.

Unfortunately, Berdan was faced with determined opposition from the head of the US Ordnance, James W. Ripley. While the battle to secure the Sharps continued, some of the Sharpshooters were issued with the Colt rifle, an unusual design based on Samuel Colt's revolver and capable of firing five shots without reloading. Accuracy was limited but the rate of fire was undoubtedly impressive. The Sharpshooters were undecided on the merits of the Colt. Charles Stevens described an early action during the Peninsular Campaign, where Colts were used by the Sharpshooters:

> Advancing ahead they met the Confederate fire with their five-shooters in a manner that evidently surprised the foe, who little expected such rapid firing. The revolving chambers of the Colts were soon heated up, and right here a most favourable opportunity was presented to test these heretofore doubtful arms. The boys were compelled to admit that they were not so bad after all, having done good work with them.[7]

Other Sharpshooters were less keen, preferring a more traditional firearm. Rudolf Aschmann summed up the prevailing attitude towards the Colt: 'Our Colt guns were a fairly good weapon in skilled hands, only their construction was a little too complicated, and it was hard for the soldier in the field to keep them in good working order, particularly in rainy weather.'[8]

Whatever the merits of the Colt, some 2000 Sharps breech-loaders were eventually issued to the Sharpshooters in the spring of 1862, and from then on became their standard firearm throughout the conflict. The arrival of the Sharps was popular with the men, as Aschmann observed:

> The new rifles were all that we could have wished for. Besides being easy and quick to load in any position, they fired accurately even at great distances. They were easy to clean and keep in good working order, and more than any other gun in the army they had the look of a weapon worthy of a sharpshooter. They left nothing to be desired in the soundness of quality, and soon this rifle came to be so well liked in the regiment that even companies which were equipped with target rifles exchanged the latter for the new guns.[9]

Despite Aschmann's comment, the target rifle had sufficient specialist uses to ensure a limited issue to selected individuals within the regiment. Sergeant Wyman S. White of the 2nd Regiment of Sharpshooters provided this information on the use of target rifles:

> Heavy target rifles with telescopic sights were used throughout the war for 'Special Sharpshooting'. There were not many of them, but they were assigned to those soldiers who were considered the best shots. Each gun had a special wooden case, and when the unit moved, it was carried by a special wagon. When the man put his telescopic rifle away, he took up his Sharps rifle again, and moved with the troops until a special duty required the use of his long-range rifle again. He oftimes operated as an independent marksman in various parts of the line where he thought he could do the most good.[10]

In the Confederate states, target shooting was less well developed, but the high overall quality of Southern marksmanship compensated for this. All infantry units employed large numbers of skirmishers to fight in front of the main line; the Confederates also deployed specially armed marksmen to operate in a sniper role. The target-rifle Union marksmen were confined to a few specialized units such as the Sharpshooters, but the Confederates' were spread throughout the Army.

The Confederate marsksmen were usually equipped with high-quality weapons imported from abroad, the most common being the British-manufactured Whitworth and Kerr rifles. These were not civilian target rifles, but military weapons capable of the highest standards of accuracy. All Whitworths were manufactured to take a Davison telescopic sight, a reflection of the intended use for long-range shooting. Of the Kerr rifle, Ed Thompson, a soldier in the famous Kentucky Orphan Brigade, wrote:

> [It] was a long-range muzzle-loading rifle that would kill at a distance of a mile or more, requiring a particular powder; and there was some difficulty in charging it, so that it was not likely to be fully effective except in the hands of a cool and composed man. The use of ordinary powder made it necessary to swab out the barrel after every fourth or fifth shot. There was prolonged target practice in which a number of men engaged, and from those who had proved to be the best shots, and were known to be otherwise thoroughly reliable, ten were finally chosen.[11]

The Orphan Brigade was composed of the 2nd, 4th, 5th, 6th and 9th Kentucky Infantry Regiments, plus a battery of artillery. Not only were the Kentucky Confederates forced to fight outside their home state (Kentucky being a border state which officially stayed in the Union), they lost a succession of commanding officers, and consequently received the nickname the 'Orphan Brigade'. One of the most distinguished formations within the Confederate Army, they were noted for the quality of their sharpshooters.

The limited numbers of Whitworth and Kerr rifles available to the Confederacy ensured that they were assigned to only the best men. Competitions to secure them were held in the Southern states, and the prize eagerly sought by rifle devotees. Sam Watkins of the 1st Tennessee Regiment described one such competition in Dalton, Georgia:

> By some hook, or crook, or blockade running or smuggling, or something of the sort, the Confederate States government had come into possession of a small number of Whitworth guns, the finest long-range guns in the world, and a monopoly by the English government. They were to be given to the best shots in the army.
>
> One day Captain Joe P. Lee and Company H went out to shoot at a target for the gun. We all wanted to shoot the gun, because if we got it, we would be sharpshooters, and be relieved from camp duty, etc. All the generals and officers came out to see us shoot. The mark was put up about five hundred yards on a hill, and each of us had three shots. Every shot that was fired hit the board, but there was one man who came a little closer to the spot than any other one, and the Whitworth was awarded him.
>
> As we just turned round to go back to camp, a buck rabbit jumped up, and was streaking it as fast as he could, when Jimmy Webster raised his gun and pulled down on him, and cut the rabbit's head entirely off with a minnie [Minié] ball right back off his ears. He was about two hundred and fifty yards off. It might have been an accidental shot, but General Leonidas Polk laughed very heartily at the incident, and I heard him ask one of his staff if the Whitworth gun had been awarded. The staff officer responded that it had, and that a certain man in Farquharson's regiment – the Fourth Tennessee – was the successful contestant, and I heard General Polk remark, 'I wish I had another gun to give, I would give it to the young man that shot the rabbit's head off.'[12]

The test of any soldier's training was combat, and in at least one instance, a Confederate officer would take his recruits on to the battlefield to see how they measured up. After an 'English admirer' had donated a dozen Kerr rifles to the Orphan Brigade, its then commander, General Lewis, formed a special sharpshooter company comprising the two best marksmen from each regiment. The company was placed under the command of Lieutenant George H. Burton, a tough, battle-hardened officer. William C. Davis, a present-day historian of Kentucky's involvement in the Civil War, describes the Orphan approach to training and tactics:

> In the campaign to come [in Georgia], Burton personally tested the recruit's grit by conducting him to a heavy artillery fire and standing in it with him. Lewis instructed the company never to approach within four hundred yards of the enemy, but rather to keep their distance and use their superior rifles to bring down federal artillerists and officers. They were to work their way close to the enemy at night, spot his artillery positions, and then silence the batteries after dawn if possible. So prized did membership in this band become that when one of Burton's men was killed – as many men were – there were numbers anxious to take his place.[13]

The Confederate sharpshooters fought individually, dispersed over a wide front and given freedom to roam and choose the targets they considered most appropriate. Some idea of the sophisticated tactical awareness of the Confederate sharpshooter can be gleaned from this account by John West, one of 13 Whitworth-armed Georgians fighting in General Robert E. Lee's Army of Northern Virginia:

> We frequently resorted to various artifices in our warfare. Sometimes we would climb a tree and pin leaves all over our clothes to keep their color from betraying us. When two of us would be together and a Yankee sharpshooter would be trying to get a shot at us, one of us would put his hat on a ramrod and poke it up from behind the object that concealed

and protected us, and when the Yankee showed his head to shoot at the hat the other one would put a bullet through his head. I have shot them out of trees and seen them fall like 'coons. When we were in grass or grain we would fire and fall over and roll several yards from the spot whence we fired and the Yankee sharpshooters would fire away at the smoke.

Artillerymen could stand anything better than they could sharpshooting, and they would turn their guns upon a sharpshooter as quick as they would upon a battery. You see, we could pick off the gunners so easily. Myself and a comrade completely silenced a battery of six guns in less than two hours on one occasion. The battery was then stormed and captured. I heard General Lee say he would rather have those thirteen sharpshooters than any regiment in the army.[14]

West was undoubtedly correct in identifying the particular problems faced by artillery in the Civil War. In the Napoleonic period, gunners could usually keep infantry at bay with canister and grape shot, and by 1900 artillery had the range to operate well behind the front line, but in the 1860s sharpshooters outranged artillery canister shot. As a result, gunners became fatally vulnerable to marksmen. Sam Watkins, the infantryman from Tennessee, contributed this exclamatory narrative of operations against Union artillery in Georgia:

We form line of battle on top of Rocky Face Ridge, and here we are face to face with the enemy. Why don't you unbottle your thunderbolts and dash us to pieces? Ha! here it comes; the boom of a cannon and the bursting of a shell in our midst. Ha! Ha! give us another blizzard! Boom! Boom! That's all right, you ain't hurting nothing! 'Hold on boys,' says a sharpshooter armed with a Whitworth gun, 'I'll stop that racket. Wait until I see her smoke again!' Boom! Boom! the keen crack of the Whitworth rings upon the frosty morning air; the cannoneers are seen to lie down; something is going on. 'Yes, yonder is a fellow being carried off in a litter.' Bang! Bang! goes the Whitworth, and the battery is seen to limber off to the rear.[15]

At times sharpshooters would work with their own gunners to defeat those of the enemy. Johnny Green, another soldier in the Kentucky Orphan Brigade, was part of the rearguard attempting to slow Sherman's advance through Georgia:

> The enemy came at us pretty early. We had the advantage of position, having posted our battery so it would sweep that road for nearly a mile. We hid our sharpshooters also in the edge of the swamp and when their battery would reply to ours those keen-eyed marksmen would pick off their artillerymen, so that our battery had much the best of the fight. This kept up nearly all day when they made a final dash and we fell back, but we punished them so severely that they did not follow us far that day.[16]

The sharpshooters' war against artillerymen was only part of a wider strategy of eliminating high-value targets. On both sides, many soldiers of general rank became casualties as a result of sharpshooter fire; of these the most famous was the death of a Union corps commander, General John Sedgwick, at the Battle of Spotsylvania on 9 May 1864. While rallying his troops, who had been unnerved by Confederate rifle fire, Sedgwick good-naturedly chided them when they flinched at the sound of bullets flying overhead. His famous last words were: 'Why boys, they couldn't hit an elephant at this distance . . .' With that he fell dying, hit in the face by a sharpshooter's bullet.

Of course, generals and their mounted staffs presented large and highly visible targets for sharpshooters, and at times they must have seemed hard to miss. One of the most exceptional feats of long-range shooting was directed against such a group of Union officers, a story later recounted by Captain F.S. Harris:

I remember a shot by a Tennessee lieutenant in 1864, which I have never seen equalled. Soon after Grant's mine exploded near Petersburg in the summer of 1864, an officer in Archer's Tennessee Brigade observed a party of horsemen ascend an eminence far in the rear of the Federal lines. He called Captain Slade, Chief Engineer of A.P. Hill's Corps, who was passing at that moment, and asked him to calculate the distance. Slade estimated it to be 2250 yards. Just as one of the men, apparently a general, rode away from the group and stopped at the highest point, the lieutenant took a Whitworth rifle belonging to one of the sharpshooters in that Brigade, trained the gun on him with globe sight, deliberately aimed and fired.

The officer fell from his horse, and his staff gathered around him quickly. Two more shots were fired in rapid succession, and three men were carried from that place. A few days later a Northern paper announced that General ——, I forget the name, and several of his staff were killed by Rebel sharpshooters at long range.[17]

Assuming that the distance in the account was accurate, and even allowing for the quality of the Whitworth rifle and the size of the target, this remains an amazing feat of marksmanship.

Among other Union losses were those of General John F. Reynolds, shot by a sharpshooter at Gettysburg, and Brigadier General William H. Lyttle, who was killed at Chickamauga. Although there is some speculation about the cause of Lyttle's death, there can be no doubt that the loss of this gallant soldier was a grievous setback for the Union forces. According to one witness, Major John N. Edwards, Lyttle was brought down by a Whitworth shot:

In his flanks was a company of skirmishers armed with the Whitworth rifles, and, fortunately, not ten rods away a lieutenant of this company was operating with a dozen

marksmen. Hindman called him up, ordered him to fire
upon the Federal commander, and kill him if possible, well
knowing the effect of his death upon the men. Coolly as if
on dress parade, the young officer stepped out with his men
to the front and took deliberate aim under a galling fire.
Twelve rifles cracked simultaneously. Rider and steed went
down together, and the black mane of the horse waved over
Lyttle. Three bullets struck him, seven his horse, a wonderful
fire and remarkable for terrible accuracy.[18]

The killing of high-ranking officers was not an exclusively Confederate speciality, and in their turn Southern generals fell
victim to Northern sharpshooters. In one instance, during the
battle at Pleasant Hill, Louisiana, Captain John H. Metcalf was
credited with killing a Confederate general at a range of 'one mile
and 187 feet', using a target rifle (although it should be noted
that some doubt remains over the veracity of this instance of long-
range shooting).

Another target highly prized by sharpshooters was the killing
of other sharpshooters on the enemy side. Many stories detail the
lengths taken to ensure a successful outcome. One such account
by William C. Davis describes operations on the Kennesaw line
during June 1864, as the Orphan Brigade fought to hold its
position against superior Union odds:

The skirmishing had been hard that day, and as men died on
the line, others volunteered to take their places. Sergeant
Tom Cox, though not on the duty list, asked to take a
comrade's place as a replacement, only to fall to a sharpshooter's bullet. When an orderly arrived to ask another man to
come to the skirmishing line, Virginius Hutchen – the same
man who so cursed his officers two years before that he was
arrested for inciting mutiny – stepped forward calmly. 'My
time has nearly come,' he said with resignation, then went
to the line. He saw several of his fellow Orphans, each with

a bullet hole in the centre of the forehead, testament to the deadly skill of the enemy sharpshooters. He asked where the dead men had been in the line, and was shown a pile of stones beside a tree. He decided to shift things a bit and took his ground several feet away from the barricades.

Shortly he crawled to the rocks, put his hat atop a stick, and held it above the breastwork. Instantly a bullet perforated his headgear. Three more times he repeated this charade until he spotted a tuft of smoke that betrayed the enemy marksman's position. At that moment Taylor McCoy of the brigade sharpshooters approached, and Hutchen borrowed his Kerr rifle. One shot brought down the enemy picket and Hutchen felt well satisfied in having, however slightly, evened the score for his departed comrades.[19]

Another sharpshooter duel, conducted during the final stages of the war at Hare House, is described by Dan Flores:

Here, firing through tiny slits in the trench breastworks, a Southern sharpshooter using a scoped rifle (probably a Whitworth) and James Ragin, a Berdan Sharpshooter who doted on his own scoped target rifle, conducted a private war of their own. The Rebel sniper had already driven one man from the slit with his pin-point shooting, but Ragin proposed to see the affair right through to the end. Eventually, but not before a numerous exchange of near-misses had occurred, the inevitable happened. Both sharpshooters touched their hair triggers simultaneously, and the two rifles, several hundreds of yards apart, boomed as one. As Ragin's rifle recoiled he was conscious of a searing pain in the head; the bullet from the Confederate rifle had clipped his hair at the scalp, coming within a half-inch of killing him instantly.

From the Southern firing slit came no more shots, no more white puffs of smoke. For at least one Confederate sharpshooter the War for Southern Independence was over.[20]

On active service sharpshooters lived a dangerous life; for extended periods they fought in the front line, and sometimes beyond it. Patrols and other reconnaissance actions ensured that they were often in close contact with the enemy, even during otherwise quiet periods. Skirmish actions were a constant feature of life, and contributed to the draining losses suffered by sharpshooter units. These small-scale battles were often confused affairs, where quick thinking could make the difference between life and death. The Berdan Sharpshooter, Rudolf Aschmann, describes one such engagement, just before the battle at Spotsylvania Courthouse. Aschmann, along with a small patrol including the regimental adjutant, pressed forward beyond their own lines and spotted a lone rider on patrol:

> Not knowing whether perhaps he was a deserter I called out to him to come near, whereupon he immediately aimed his carbine at us and pulled the trigger. But luckily the shot misfired. My two companions quickly shot off their rifles, the adjutant galloped off calling to the men to fall back, and in an instant I found myself alone, not quite knowing how.
>
> When I looked back at the enemy once more I noticed that a detachment of thirty horsemen were charging at full speed across a field towards the road, while the patroller about to gallop down the road shouted a thunderous 'Halt, or I shoot!' But now it was my turn to shoot. My rifle went off and I saw the man reel in his saddle. Two comrades rushed to him, one of them taking the horse's reins and the other catching the injured man in his arms. Then I began to run and soon reached my company which was just getting ready to come to my assistance.[21]

Sharpshooters could be particularly deadly in ambushes, surprising a complacent enemy who was not expecting to come under long-range rifle fire. *The Illustrated London News* had dispatched the correspondent and artist Frank Vizetelly to record the war from a

Southern perspective. He cabled this report of Confederate sharp-shooters in action:

> Since the Federal Army has sought refuge in its entrench-ments at Chatanooga the object of the Confederates has been to disturb its communications, and, if possible, to break up its supply trains. With a view to this a small force of picked men from General Longstreet's corps, armed with Whitworth telescopic rifles, were posted among the crags of Racoon Mountain, overlooking the road on the other side of the Tennessee river, in use by the Federals. The position chosen was twelve miles in the rear of the enemy's works, and, unfortunately for them, was not guarded.
>
> At the invitation of Captain Gorree, who was to take charge of the party, I accompanied the expedition; and the only way of reaching the place and avoiding the Federal scouts was by taking the Indian trails through the forest heights. Scarcely had we arrived at our destination and got the men hidden behind the rocks, when, far up the gorge, could be heard distinctly the rumbling of approaching wheels. Presently emerged the leading files of an infantry escort, followed by the first wagon.
>
> The order was not to fire until the open space of the road in front of us was filled. At last came the word, and a score of sharp cracks leapt in deafening echoes from crag to crag, and a score of mules, with their riders, went down, creating the most dire confusion. There came such a stampede as was never before seen. The teams in front endeavoured to gallop on, but were prevented by one or more mules hanging dead in the traces; others behind made an effort to turn, but got blocked by those in the rear. Crack! Crack! still went the rifles, until the road was choked with dead and dying men and mules, and overturned wagons. The escort, after firing a few shots in return, fled panic-stricken, followed by the exulting shouts of the Confederate riflemen.[22]

A diary entry from a Union soldier outlines the experience of coming under sudden fire from long-range sharpshooters, although in this instance the extreme range prevented any fatalities. Private Rice C. Bull, of the 123rd New York Infantry, and his friend Nat Rowell were taking a stroll behind the lines, while their regiment was acting as part of a blocking force southwest of Marietta, Georgia:

> An opening in the road furnished a good range for their sharpshooters; not knowing this we walked some distance along this road. We were soon greeted with wheezy, long-drawn-out sounds made by nearly spent bullets that were going slower than the sound was coming to us. It seemed that the Johnnies with their English rifles had the range but it was a long one. Before we could get to the woods at the side of the road a spent bullet struck Nat on the instep. We got the shoe off and found it was only a bad bruise. His foot swelled so quick he could not get the shoe back and I had a slow time getting him back to camp. He was lame for several days.[23]

While sharpshooters had been trained for open-order skirmishing, they were equally useful in siege warfare. The American Civil War witnessed long-drawn-out battles for fixed positions, involving the extensive use of breastworks and entrenchments. As the war progressed, the primacy of firepower over manoeuvre made trench-bound fighting increasingly prevalent, although this was apparent as early as 1862, at the siege of Yorktown during the Peninsular Campaign. Rudolf Aschmann was present at this Campaign, and his description of the fighting presents striking parallels with the Western Front in the First World War:

> Our service here was very unpleasant and extremely dangerous. On the wing, the two armies were separated by only the Warwick Creek, a sluggish, swampy water. The Confederates' side of the water was strongly fortified, and just at the

place where a dam made a crossing possible, the enemy had mounted a battery and dug rifle pits which were a great source of harassment for out artillery as well as for outposts.

In two consecutive nights we dug a trench, reinforced it with fascines and gabions and equipped it with well-placed loopholes. It was not a task for the fainthearted: working in pitch-dark night without lights of any kind, silent and avoiding all noises as much as possible, at less than 250 steps from the enemy. The slightest noise betraying us to the enemy would have attracted the fire of eight cannon, and there was no tree, no rock, no hillock which could have afforded any protection. Even when the trench was finished, its short distance from enemy marksmen made entering or leaving it impossible during daylight. The detachments which were ordered there always had to enter the shelter before daybreak and remain all day long until the darkness of the night enabled them to withdraw.

From this ditch, about six feet deep, we closely watched the enemy cannoneers and marksmen and very soon made it impossible for the former to serve their guns. But our opponent, too, had his watchful eyes upon us. It was dangerous to show one's head above the breastwork for even an instant, and more than one man paid for this carelessness with his life. Moreover, this duty was extremely unpleasant. Each detachment had to stay in the trench for sixteen hours. This would have been all right in dry weather. But the incessant rain filled the trench with water in which we had to stand up to our knees all day long, because mud and sludge made it impossible to either sit or lie down. But good results demand sacrifices, and we carried out our task in a manner which brought us the praise even of our generals.

The enemy marksmen could no longer continue unchecked and the artillery had to quit firing, for woe to him who showed himself for an instant. His life was threatened by a dozen well-aimed bullets. The men who worked in the rear of us could safely construct ramparts and

entrenchments, and soon our enemies were surprised by full
rounds of artillery fire.[24]

The Peninsular campaign revealed many shortcomings in Union
generalship, but Berdan's Sharpshooters made a reputation as
skilled marksmen who were capable of doing serious damage to
the enemy. Although writing from the perspective of a Sharpshooter himself, William Ripley was able to recount that, 'Gun
after gun was silenced and abandoned, until within an hour every
embrasure within a range of a thousand yards was silent. The
rebel infantry, which at first responded with a vigorous fire, found
that exposure of a head meant grave danger, if not death.'[25]

During the summer of 1863, Union forces laid siege to
Battery Wagner, a defensive position on one of the small islands
guarding Charleston, South Carolina. The confrontation between
Confederate sharpshooters and Federal artillery and siege engineers
continued unabated for months. Captain S.A. Asche of the
Confederate Ordnance Department described his role in deploying
Whitworth rifles:

> I had been instrumental in having our agent in England,
> Major Huse, obtain for us some telescopes for rifles, and he
> sent besides half a dozen Whitworth rifles with telescopes
> afixed. Two were given me. I had some sandbags removed
> from a wall, leaving two holes, at each of which a marksman
> with a Whitworth rifle stood ready to fire. A hat raised on a
> ramrod drew the fire of some Federal sharpshooter, who
> would then look to see the effect — and quick would come
> the Whitworth bullet. Those guns were fatal at fifteen
> hundred yards.[26]

Union engineers were forced to work under the cover of darkness,
and devote much of their energy to constructing bullet-proof
defences as protection against the Confederate sharpshooters.
Despite their best efforts, the Southern artillery and marksmen

continued to cause casualties. A Union engineer, Major Thomas B. Brooks, made this report to a superior officer on 22 August:

> At 11.30 o'clock the work was discontinued on account of this fire and that of the enemy's sharpshooters, who occupied a position in the ruins on the left, which enfilade the trench. The Eighty-Fifth Pennsylvania did the advance guard duty yesterday. Their casualties amounted, I am informed, to twenty-five during the tour of duty.[27]

Early in September the garrison of Battery Wagner – faced by overwhelming numbers – knew defeat was inevitable. To avoid capture the Confederate troops executed a skilful evacuation of their position, their withdrawal covered by the ubiquitous sharpshooters.

Despite the heavy casualties suffered by sharpshooters on both sides, there were lighter moments which broke the monotony and grimness of the war. John S. Jackman of the Orphan Brigade was involved in some 'light' skirmishing during the retreat of Confederate forces in the face of Sherman's march into Georgia: 'Went with company on the skirmish line. We had a lazy time all day. After taking a good nap, I finished the day reading Miss Evans' new novel, *Micaria*. Showery all day. Continuous cannonading in front.'[28]

A few days later Jackman had finished Miss Evans' novel – thinking it 'of not much force' – and was engaged in a further skirmish with Union infantry:

> The Federal sharpshooters are very close to us, and keep a stream of bullets coming over the hill all the time. There were some old smooth-bore muskets left lying around by the Tenn. troops, also a box of buck-and-ball cartridges, and our boys have been trying to see who could shoot the largest loads out of them, directing the fire down through the woods, at the Federal sharpshooters. Some have shot a

handful of buckshot, and several balls, at a single load. Sharpshooting kept up briskly from both sides.[29]

On the Union side, Rudolf Aschmann describes a similar absence of serious military endeavour during a pause in the three-day Battle of the Wilderness:

> During the afternoon the fire slackened on both sides, and soon the men even began to amuse themselves by shooting at one another not with bullets but with ramrods from the piles of shotguns that lay scattered about. These missiles cut through the air with a hissing sound not that unlike that of a bomb and were received on both sides with shouts of laughter.[30]

Fraternization between the two sides was not unusual. At the siege of Petersburg, Aschmann explained some of the workings of the North–South 'live and let live' system:

> For a while the guards at our front were on the best of terms with each other. I shall prove this with a few examples. One day when we were on picket duty a shot fell from the Confederate side without, however, striking anyone on our side. Indignant at this breach of the peace we made ready to shoot back but the enemy immediately gave a sign that this disturbance was due to a mistake. Soon after, the delinquent appeared in front of the enemy line where as a penalty he had to parade back and forth for two hours in plain view of both picket teams, carrying a heavy beam on his shoulder. Loud shouts of applause indicated to the Southerners that this was satisfaction enough and that peace had now been completely restored. Occasionally the higher command of one or the other side would give orders to open fire on the opposite patrols. In such case they usually gave each other a warning signal, and once we even heard a Southerner call: 'Yankees, lie down! We have orders to shoot!' Immediately fire was opened and kept up all day with intensity.[31]

Sharpshooters generally went about their deadly business with few
moral qualms, but at times soldiers were troubled by the suffering
they caused. Regret of this nature was displayed by a Confederate
soldier returning from picket-line duties in a dejected mood.
When asked by his comrades if all was well, he exclaimed: 'I did
not want to kill the fellow. I struck him, and he screamed. It was
the cry of a boy! I don't like to think of having killed a boy!'

There was a general feeling among other troops that the
sharpshooter did not 'play fair', and in some way violated the
unwritten rules of war. One Northerner complained that sharpshooters would 'sneak around trees or lurk behind stumps' and
from this vantage point 'murder a few men'.

J.W. Minnich, a trooper in a Georgia cavalry regiment,
observed the moral ambivalence felt by sharpshooters in the front
line. Minnich had teamed up with a Whitworth rifleman engaged
in some between-the-lines skirmishing. After a while they saw a
blue-clad horsemen calmly surveying Confederate positions. Minnich describes the subsequent course of events:

After some minutes of close observation of our friend 'soldier
on a horse' – all the while keeping a sharp lookout to the
front – my friend with the long-range Whitworth finally
spoke, more to himself than to me, saying, 'I have a good
mind to unhorse that fellow'.

'Can you do it?' asked I.

'I think I can. What distance do you make it?'

I was a tolerably good judge of distance in those days,
having had considerable practice, and was pretty sure I'd
make a miss of it with my Enfield. I told him I judged the
distance to be about seven hundred and fifty to eight hundred
yards. He adjusted his sights. 'I'll put them to eight hundred
and if I don't miss my distance, I'll take him through the
hips.'

With my gun resting butt on the ground, and intently
watching the vidette to note the effect of my friend's shot, I
held my breath. My companion raised his rifle slowly and

deliberately 'offhand', as if aiming at some inanimate target and pulled the trigger.

It was one of the prettiest shots I have ever seen. Or even heard of. But – and here fate, destiny, or whatever we may term it – Providence intervened. I clearly saw the vidette suddenly dig his spur into the horse's flank and make one step forward, when the bullet struck the horse on the rump, raising a little cloud of dust. It did not strike low enough to disable him in any way, as the violent switching of his tail proved, but caused him to make a startled leap forward that nearly unseated his rider, who, no doubt was a bit startled himself, for he wheeled to the left and back into the woods out of sight.

My companion dropped the butt of his rifle to the ground and drew out another cartridge to reload as we watched the blue coat take to the 'tall timbers'. 'By golly! I missed him,' were his first words, 'and I was sure I'd get him.' And after a moment he added: 'Well, his time had not come to die today, and I am not sorry he got away. It looked too much like murder. He was a brave fellow, and I hope he'll get through all right.'

To tell the truth, I felt a bit like that myself, although in those days I felt it my duty to do the enemy as much damage as I could, and never failed to try when the opportunity presented.[32]

Sharpshooters were especially hardened to the rigours of war, and, in contrast to the infantry of the line, tended to be men of an independent cast of mind. There were a number of sharpshooter 'characters', the most famous being 'California Joe', an early member of Berdan's Sharpshooters. His real name was Truman Head, although he was universally known by his nickname after time spent in the California Gold Rush. His popularity among the regiment was underlined by this comment from Charles Stevens: 'Joe was one of those splendid characters that made him a hero in spite of himself. Entirely free from brag and bluster, Joe was an unassuming man, past middle age, short in stature, light in weight, and a true gentleman in every sense of the word. He was always a special favourite with the entire command.'[33] Another contemporary described him as having 'an eye as keen as a hawk, nerves as steady as can be, and an endowment of hair and whiskers Rubens would have liked for a patriarchal portrait.'[34]

The Confederates also had men they considered to be a cut above the average. While guarding Confederate prisoners immediately after the end of the war, Corporal James K. Hosmer met a renowned marksman. Hosmer recounts a conversation with a group of Confederates:

> 'Here comes Old Thous'n' Yards!' said they, as a broad tall Arkansan, with a beard heavy as Spanish moss on an oak, and a quick, dark eye, lank frame, came swinging down from the parapet. All the Rebels make way for him with deference. He was 'Old Thous'n' Yards' with everyone, and turned out to be the great sharpshooter of that part of the works. I inquired about him and found he was a famous backwoods man and hunter, who, with a proper rifle, was really sure of a bear or buffalo at the distance of a thousand yards.
>
> I borrowed a chew of tobacco, and won the perennial friendship of 'Old Thous'n' Yards' by bestowing it upon him. I fear more than one of our poor fellows has felt for his skin; but, for all that, he was a good-natured fellow, with a fine frame and noble countenance, a physique to whose vigor and masculine beauty, prairies and mountain paths and wild chases had contributed.[35]

In a period when the rifle was an unusually prominent weapon, the role of sharpshooters was undoubtedly important. The sharpshooters had a high opinion of their own merits, but it was one generally shared by others – on both sides. Following the fine performance by Berdan's Sharpshooters at the siege of Yorktown, they were commended by a staff officer: 'The Commanding General instructs me to say to you that he is glad to learn, from the admissions of the enemy themselves, that they begin to fear your Sharpshooters. Your men have caused a number of the rebels to bite the dust.' This view was supported by a captured Confederate officer, who said: 'The Sharpshooters were the worst men we had to contend with'.

Berdan's Sharpshooters were fortunate in being armed with the superior Sharps rifle; two full battalions of Sharps-armed troops had obvious advantages over men equipped with conventional Enfield or Springfield rifle-muskets. This factor was acknowledged by an officer in the regiment, following the success at Chancellorsville:

> The complete and almost instant success of a mere line of skirmishers in turning back the enemy approaching in force I confess surprised me. How much of that success is to be attributed to the superior skill of these skirmishers as marksmen, how much to the rapidity with which their Sharps rifles can be loaded and fired, [and] how much to the remarkable coolness and steadiness of the men themselves, I do not pretend to know.[36]

The Confederate soldier John S. Jackman wrote: 'We think the Federal Sharpshooters are 'terribly' inconvenient, and they have the same opinion of ours. We captured one of them, and he told Gen'l Bate that our sharpshooting is excellent, but that our artillery is not worth a d———n.'[37]

The quality of Southern sharpshooters, particularly those armed with Whitworth and Kerr rifles, was of the highest measure. Contemporary testimony to their prowess was a matter of record. Describing the sharpshooters of Mahone's Brigade, Captain John E. Laughton wrote: 'Probably the most effective troops in the late civil war, for the number of men engaged, were the sharpshooters.' One Southern veteran summed up the contribution of the Confederate sharpshooters with a few heartfelt words: 'The Whitworth men on the Confederate side were a class quite to themselves.'[38]

Chapter Four

The First World War

The opening phase of the First World War, in which armies fought each other in open, mobile warfare, lasted less than two months. From October 1914 onwards, trench systems of increasing complexity were dug across Europe. This fixed warfare, dominated by firepower, came as a shock to the military mind, and soldiers on both sides struggled to come to terms with new ways of waging war. But from the outset, trench warfare presented opportunities for sniping on a massive scale.

The close proximity of the opposing forces enabled enterprising and competent shots to pick off unwary troops on the other side of no man's land with comparative ease. As the enemy became more wary, however, so the sniper's task became more difficult. The British big-game-hunter-turned-sniper, Major Herbert Hesketh-Prichard, explained how the minimal size and fleeting nature of enemy targets necessitated the highest standards of rifle fire:

When we settled down to trench warfare, and the most skilful might spend a month in the trenches without ever seeing, except perhaps at dawn, the whole of a German, and when during the day one got but a glimpse or two of the troglodytic enemy, there arose the need for very accurate shooting. The mark was often but a head or half a face, or a loophole behind which lurked a German sniper, and no sighting shot was possible because it 'put down the target'. The smallest of big game animals did not present so small a mark as the German head, so that sniping became the highest and most difficult of all forms of rifle shooting.[1]

Despite the efforts of a few devotees such as Hesketh-Prichard, the British were not prepared for sniping. In contrast, the Germans grasped at it as a means to inflict heavy and demoralizing casualties on their opponents. German superiority in sniping lasted throughout 1914 and 1915, and only in the following year did the British establish any real parity. Hesketh-Prichard was well aware of the early German dominance:

> At this time [1915] the skill of the German sniper had become a byword, and in the early days of trench warfare brave German riflemen used to lie between the lines, sending their bullets through the head of any officer or man who dared to look over our parapet. These Germans, who were often forest guards, and sometimes battle police, did their business with a skill and a gallantry which must be freely acknowledged. From the ruined house or the field of decaying roots, sometimes resting their rifles on the bodies of the dead, they sent forth a plague of head-shot wounds into the British line. Their marks were small, but when they hit they usually killed their man, and the hardiest soldier turned sick when he saw the effect of the pointed German bullet, which was apt to keyhole so that the little hole in the forehead where it entered often became a huge tear, the size of a man's fist, on the other side of the stricken man's head.[2]

Many other British soldiers testified to the shooting prowess of their opponents. Frank Richards, an Old Contemptible who served in the Royal Welch Fusiliers throughout the war, wrote a classic account of life in the trenches. In the winter of 1914, he recorded this example of German marksmanship:

> We used to fix iron plates in the parapet of our trench, concealing them as cleverly as we could, but in a day or two

the enemy generally discovered them and would rattle bullets all around. Later they had a bullet that would penetrate them; a man was never safe even when firing behind one of these plates. The hole in the iron plate was just large enough to put the muzzle of the rifle through, and one morning I saw the greatest shooting feat that I ever saw during the whole of the war. A man named Blacktin was firing from behind one of these plates. He had fired two rounds and was just about to pull the trigger to fire the third when he seemed to be hurled against the back of the trench, the rifle falling from his hands. A German sniper had fired and his bullet had entered the barrel of Blacktin's rifle, where it was now lodged fast, splitting the top end of the barrel the same way as a man would peel a banana. Blacktin's shoulder was badly bruised.[3]

Harold Stainton, a volunteer in the Liverpool Scottish, attempted to redress the balance in a contest with a German sniper in December 1914:

Hereabouts a sniper fell to my rifle. In the first gleam of half-light before dawn when the trees of the Petit Bois were emerging from silhouette into things with shape and form, I was peeping over my sights through the base of a hedge when a large form silently rose and obscured my sight of the trees. It could only have been an enemy, and instinctively I took a quick rough bead on the massive target and pulled the trigger. Whatever it was sank to the ground and out of my vision. Now I was tense with suppressed excitement. I felt sure I had disposed of a murderous sniper and in a bated whisper told my neighbour so – but where was the evidence? I had to wait impatiently for dawn to break before finding it: the grey uniformed body of a big and bulky German lying silent and still, three or four yards away.[4]

Ominously, Stainton's manuscript ends abruptly, midway through
the next sentence. Other British soldiers engaging German snipers
regularly came off worse in sniping duels. Frank Richards
described the death of two comrades during the latter stages of
the First Battle of Ypres (November 1914) when a Private Miles
decided to try his hand at sniping: 'Miles had just claimed to have
popped a German over when he got a bullet through the head
himself. The same evening Corporal Pardoe also got killed in the
same way.'[5]

Even when the British attempted to organize their snipers and
scouts on a more professional basis, they continued to experience
setbacks. Major F.M. Crum established a reputation as a renowned
sniper trainer, but initially, he had many problems. Stationed in
the village of Blagny on the Arras Front, he found the Germans
particularly dangerous:

We started badly. The very first day, I went down to explore
with Corporals Otterwell and Riches, my two best men. We
climbed every ruined house at the back, which seemed to
offer a view, and so, after a thorough search, worked our way
forward to the foremost trenches and listening posts. Everything seemed still, and one might have thought the Hun
had gone.

At a point where the main street ran through the village,
the cobble stones had been pulled up, and a shallow trench
dug to connect the foremost posts on each side of the road.
A few sandbags increased the amount of shelter, but still a
man had to keep down very low to avoid being seen from
the opposite barrier fifty yards away. Peeping over gradually,
I got a general view, and then proceeded to search the
opposite breastwork methodically with my telescope for
hidden loopholes. Suddenly it gave me a bit of a turn to see
the silver outline and black centre of a rifle barrel pointing
in my direction. It seemed so close. I kept quite still. Then
slowly, to my relief, it moved away from me. Keeping the
glass steady, after noting the exact position with regard to a

conspicuous pink sandbag, I slowly withdrew my head and showed it first to Otterwell and then to Riches. We all saw the rifle barrel move. Returning to cover, we considered our plans and noted the exact position on paper.

I then went back to warn the sergeant on duty that he was to caution everyone who passed to keep down, and get the trench deepened that night. As I went, a shot rang out, and I turned to see both my Corporals wounded in the bottom of the trench. Otterwell died in my arms almost immediately. So, scouting the way for the benefit of his comrades, passed away as fine a young fellow as I have seen this war. Riches was not severely wounded. From that time onwards, a special vendetta existed between the opposing snipers at Blagny.[6]

The reasons for the slow and ineffective British reaction to sniping in the trenches were a combination of official rigidity and ignorance of the value of accurate shooting – and the casual attitude towards the dangers of trench life displayed by the average British soldier. This latter factor was underlined in a letter written by J.K. Forbes. Despite his background as a biblical scholar at Aberdeen University, Forbes joined a territorial battalion in the Gordon Highlanders and became its sniper sergeant. Forbes wrote:

We British take risks the Germans do not. We dispense with communication trenches in practice, though not in theory. We prefer to walk over open ground exposed to bullets and that for hundreds of yards, rather than suffer the certain discomfort of the narrow muddy communication trenches where one must have knee-boots, or have mud and muddy water soak through puttees and socks to the very toes.[7]

Despite Forbes's strictures on his fellow countrymen, he himself was not above a degree of derring-do that can only be considered pure recklessness. After a stint of night-time observation duty, he admitted: 'I find a certain sporting fascination in exposing a full head and shoulders for two seconds during the glare [of illumination rockets], withdrawing sharply and hearing the bullets whiz. You must not do it twice in the same place though, or if you risk it twice, not a third time.'[8]

The construction of British front-line trenches also favoured the Germans. Captain D.L. Cox of the Royal Sussex Regiment explained how this affected his troops, while his battalion was in the front line near Armentières in July 1915:

The Germans lacked little in thoroughness – for instance, they very wisely came to the conclusion that the outline of their trenches should be as untidy as and irregular as possible; also that a variety of colour was advisable. Our trenches, on the other hand, were 'dressed from the right', so to speak; a stretch of sandbags in perfect alignment, with the result that any projecting object was immediately observed.

GHQ eventually came to the conclusion that a too rigid sense of order and tidiness, although perhaps pleasing to the eye, was costing a considerable number of casualties. Their next move, therefore, was to send the battalion a consignment of paint and brushes. It was naturally left to the discretion of the company commanders as to when the paint should be applied.

Two of our company, presumably decorators by trade, took kindly to the idea, and without waiting for the necessary orders, they acquired the said materials and utensils and in broad daylight proceeded to daub our trenches in full view of the astonished Germans. They were left in peace for a short time, until the enemy, presumably having had his amusement, considered that such an excellent target should not be disregarded. We were therefore awakened by a few rounds of rapid and with considerable difficulty reclaimed our two enthusiasts. They were indeed lucky, for one escaped intact, and the other one was only slightly wounded.[9]

Hesketh-Prichard had long argued for the adoption of 'untidy' trenches. In order to convince GHQ, he conducted trials in which a dummy head was raised for a few seconds above the parapets of 'tidy' and 'untidy' trenches, and then fired upon; the number of hits recorded for the former was three times greater than the latter.

Apart from a general shortage of optical instruments in Britain during the early stages of the war, a further problem was the lack of training with telescopic-sighted rifles. Although ordinary iron sights could produce excellent results, a telescopic sight was essential for sniping at longer ranges, or at the very small targets that were typical of trench warfare. Many soldiers provided with telescopic-sighted rifles had little idea how to use them. There was a general tendency to forget the need for correct alignment of sight and rifle, and to assume that the fine sight-picture provided by a telescopic sight was a guarantee of success. Such were the failings of early British sniping that many battalion commanders began to doubt its effectiveness at all.

Eventually, the morale-sapping damage caused by German snipers forced the British to develop a comprehensive programme of their own. Some sniper sections were formed from the best shots in a battalion and deployed on an ad hoc basis, but increasingly, sniping was organized at a higher level, with army sniping schools producing a succession of trained sniper officers and NCOs. Fully conversant with the many aspects of sniping, successful candidates went back to their battalions to train and organize their own sniper sections.

But before effective training could begin, the right men had to be chosen for the job. Major Crum outlined the basis on which recruits should be selected, aware of the competition from other sources for good soldiers within the battalion or company:

Men selected should be reliable, intelligent, good shots, fit, educated. We are not likely to find the ideal man. The qualifications of the perfect sniper are unlimited. Moreover, the company officer naturally does not want to part with his best men, and there are other equally important branches to be considered, such as bombing, signalling, or the Lewis guns. On the arrival of drafts the sniping officer should see if there are any likely men. With us, men who have done well with the snipers have been sent for training at the NCOs' School and promoted into various companies. If the company officers know that these men may come back as good NCOs they will be more ready to send good men.[10]

The experts were agreed that the qualities required of a good sniper would include a hunting background or instinct. Hesketh-Prichard wrote: 'The truth of the matter forced itself upon me, as I spent day after day in the trenches. What was wanted, apart from organization, was neither more nor less than the hunter spirit. The hunter spends his life in trying to outwit some difficult quarry, and the step between war and hunting is but a small one.'[11]

Herbert McBride – an American marksman and captain in the Indiana National Guard who joined the Canadian Army at the outbreak of war in 1914 'in order not to miss the fun' – emphasized the value of stalking experience:

Since we no longer have an open season on Indians, about the best way to acquire the skill to advance over varied ground without being detected is by stalking wild game. And by stalking, I mean to get right up close to the animal before taking the shot, and not merely to crawl into some position at long range from which it is possible to take a shot over open ground. Crawl, roll or push yourself forward until you are within relatively close range of the target, and learn just what sort of Indian-cunning and patience the art of stalking calls for.[12]

J.K. Forbes found that his experience of trekking through the Scottish Highlands was useful when he was ordered to train the snipers in his battalion of the Gordon Highlanders:

> Quite a number of half-forgotten and much-neglected faculties of mine seem to be coming all into use again, all those years of working with maps and wandering in Scottish wildernesses, and the faculty of looking into things through a pair of glasses, and the faculty of being able to use all these quite decently when a big strain is on your nerves, when you must either raise your head into danger or the information be lost.[13]

Almost inevitably, sniper training began in a rather haphazard manner. Herbert McBride's time at 'sniping school' consisted of an improvised, 200-yard firing range just behind the Canadian lines, where in a few hours the snipers-to-be zeroed their rifles and checked the workings of their telescopic sights. The rest of their training was on the job in the front line. Promoted to sniper sergeant, J.K. Forbes instructed his own men at battalion level, as this posthumous memoir compiled by his friends reveals:

> He picked his little band – sixteen NCOs and men – showing in his choice much foresight and acquaintance with men's character, and immediately set about training them. Starting right at the first principles of ordinary observation, he gave very efficient lessons and practice in description and recognition of targets, scouting and visual training, and always paid special attention to developing the powers of observation and memory. As time went on, studying the landscape for the purpose of finding suitable sniping posts formed part of his scheme, and at this point many days were spent on bullet-proof sniper-posts which by means of careful construction and masking were absolutely invisible to the enemy. Judging distances with great accuracy, and practice

in approaching enemy positions over comparatively open ground, involving snake-like crawling, then gave place to more advanced scouting work, which in turn was replaced by ordinary sketching of positions, and finally panorama sketching.[14]

Forbes's programme was remarkably progressive and similar to the type of syllabus set at the army sniper schools which came into being during 1916. Forbes did not live long enough to see the development of sniping in the British Army, as he was killed at the Battle of Loos on 25 September 1915.

As head of the First Army School of Sniping, Observation and Scouting (which received official status in November 1916), Major Hesketh-Prichard was determined that his graduates would receive a rounded education in the sniper's craft:

At first the greater part of our teaching dealt with sniping, but as time went on the curriculum was much extended. Map reading, intelligence work, the prismatic compass, the rangefinder, instruction on crawling, ju-jitsu and physical drill were all added. In addition to these, we had continuous demonstrations of the effects of all kinds of bullets, both British and German, on the armoured steel plates used by us and by the enemy.[15]

Sniper training was also widened in scope by Major Crum on his return to home service. Crum insisted that the whole experience should be as enjoyable as possible. As part of a varied training schedule, he encouraged the acting out of 'scenes' to improve his men's appreciation of specific sniping lessons. Of this work, he wrote:

I felt that by acting, and appealing to the imagination, by the imitation German and British trenches, with men dressed up as Huns and as British troops in France, by the use of the cinema and lantern slides combined with lecturing, by the use of black goggles for teaching a man to work in the dark, and by the introduction of ju-jitsu and special training in self-defence, hundreds of lives which were being thrown away each twenty-four hours of the war might be saved and turned to good purpose.

It was very soon found that the men showed talent in the acting and were able to reproduce the exact happenings and mishaps, often combined with much amusement. Large parties came from all directions and the thing being a novelty, soon became a popular 'show'. Two hundred and fifty young soldiers, for instance, would be sent to me before going to France. They would attend a lecture at the local cinema, illustrated with lantern slides and cinema films, and sing a marching song, the words of which were thrown on the screen; during an interval they would next march to the range where there was room to accommodate them on the grandstand overlooking the trench. Then would take place some scene from trench life, men frying bacon over a brazier, making too much smoke, cleaning their rifles and making mistakes in doing so, which have often too often proved fatal, or incautiously exposing themselves and being hit and carried away on a stretcher; these and such-like scenes, all commented on by the instructor, and acted by all hands, with their own language and jokes, whether they produced loud laughter or seriousness, left an impression which lasted far longer than any amount of ordinary instruction.[16]

The improvements in British sniper training were confirmed in this anecdote from a recruit called up in 1918. F.A.J. Taylor had been despatched to a sniper school in France:

The course was most enjoyable and we learned a lot about the new art of camouflage. We had an observation exercise when we all stood on a high sand dune and were asked to spot enemy snipers concealed in the ground around us. We all had a go but got a rude surprise when at the instructor's whistle the camouflaged chaps stood up. One or two were right under our very noses and were not spotted. By proper camouflage most of them were successfully concealed and could have popped us off.[17]

The acid test of the various sniper programmes was, of course, on the battlefield. If one of the recurring themes in British First World War narratives was the dismay caused by German sniping, then another was the satisfaction expressed by British sniper officers in turning the tables on the enemy. After the heavy casualties suffered by his battalion – which had included the death of Corporal Otterwell and the wounding of Corporal Riches – Crum was determined to overcome the German menace:

With war once declared between the snipers at Blagny we set to and brought all our past experience into play in real earnest. A good system of observation was established with a central post, connected by telephone, in the roof of a high red brick building, the shell of which was still left standing.

Hidden, camouflaged loopholes were so constructed that every part of the Hun lines could be kept under constant and close observation. Some of these posts were so near to Fritz that the work of building them took many nights. To excavate the stones the snipers, in some cases, dug them out laboriously with penknives to prevent being heard. The slightest sound drew bombs and rifle grenades and every change had to be gradual so as not to attract attention.

For a time the Hun was top dog; and, being newcomers, many casualties, as many as nine in one day, took place among our men from sniping alone. Every casualty which occurred was at once investigated; nine times out of ten it was an avoidable casualty, and steps were taken to avoid its recurrence. But the maze of ruins was so confusing that often a man would be hit standing at his post from what seemed to him to be behind, and this might well have caused the lack of confidence known to the troops as 'getting the wind up'.

I hit on the idea of drawing a plan of the trenches on a blackboard, and lecturing each company that went to this tricky sector and explaining exactly how each casualty had occurred, and how unnecessary it had been. In this way men got over the idea that the Hun sniper did anything super-human, and realized what not to do. It also gave them an interest in their work.

The Hun did not long have it all his own way. After long planning and watching, Otterwell was at last avenged by our left-handed champion, Sheehan, who killed his man and silenced all sniping from that quarter.

I remember one particular point of vantage from which we 'strafed the Hun'. We called it the 'Boiler House' for it contained the remains of the boilers of some huge ruined brewery. It was a tricky place to approach, and we had two men killed the first week. We had two bombers stationed at this point, where, peeping through a small hole in the wall, they watched for the Hun not fifteen yards away.

One day, going to this post with Riches, I discovered that, by creeping forward like a cat and pushing a small ladder up on a platform above the bombing post, it was possible to look right down on the Hun, and occasionally catch a glimpse of the sentry. As I was looking my attention was attracted to earth being shovelled up at a point in the trench about fifty yards off. Presently the shovelling ceased, and a healthy young Hun emerged in shirtsleeves, spade in hand, and smoking a cigar. He rested and leaned against the wall of the trench, looking up and staring in my direction. Fifty yards through a good Zeiss glass brings a man pretty near. It was hard to keep still and realize he did not see me. I showed him to Riches who, trembling like an eager terrier, was all for a shot.

But I wanted to learn more. As I was watching, presently

from a new direction six more Huns, carrying hods on their
backs and large blocks of concrete, came up to this spot in
file. I let them proceed till they all vanished with their loads
down into the hole from which originally the thrown-up
earth had attracted attention. What were they doing? Soon
they came out and stood in a clump. This was asking too
much! Riches fired and claimed to hit two, but the cloud of
brick dust obscured my view. I only saw signs of some such
struggle as that on the day poor Otterwell fell. I remember
clearly the expression of terror on the faces of those of them
who did escape, crouching and running past us.[18]

Crum's description underlines the importance of sound organization, patience and simple cunning. It makes an interesting
comparison with an account from Frederick Sleath, in which
German dominance is ended by improvements in British sniping.
Sleath had made detailed notes about the activities of a sniper
officer in the trenches, but on submitting them to a publisher
directly after the war, he was persuaded to rewrite them in novel
form. Apart from the stilted dialogue sections (deleted here), the
story of 'Sniper Jackson' is an authentic description of sniping in
the Great War. Sleath describes how a sniper section, under the
command of the Sniper [Sleath's term for Jackson], comes to grips
with the enemy shortly after the battalion takes up position in the
front line:

The trouble started next morning at breakfast time. Two of
the men in the act of throwing empty tins over the parapet
were shot through the head. Then the second-in-command
received a bullet in the shoulder at the head of the main
communication trench when coming up with the colonel to
go the morning round. An hour later a subaltern of A
Company was shot through the body actually while walking
in the fire-trench. A feeling of disquiet settled down on the
battalion; there was no need to issue a second warning. The
men stuck to their cover like rabbits to their warren when

the guns are near at hand, and the day saw no further casualties.

The Germans always seem to know when a relief takes place in the British trenches, and it was just as if their snipers had been holding their fire until they had measured the quality of their new opponents before conducting a vigorous offensive against the [British] sniping posts located by their observers. All day they kept up a systematic fusillade on several of the posts, and so accurate was their fire, that one or two of them had to be abandoned.

Overconfidence is a vice that will lead the best of snipers astray, and the Germans sinned grievously. They had summed up their opponents as of little account, and the lack of any reply to their shooting encouraged them in this idea. But the [British sniper] section was working strenuously. It was no longer a case of morning and afternoon watch. Every one of the snipers voluntarily remained on duty all day. The Sniper divided up the German line into narrow sectors of observation, putting a sniper in charge of each, with instructions to carry out the minutest possible scrutiny, inch by inch at a time. The plan yielded immediate results, and one after another the new sniping positions were found and noted down.

The sniper who shot the subaltern of A Company was the first to suffer. From the manner in which the casualty had occurred it was obvious the fellow must have been stationed above ground level. A clump of tall trees stood some distance behind the German line a little to the left of the point, and only from one of them could the shot have come.

It is a fairly simple matter to detect a sniper in a tree if the tree stands by itself. But where it is one of a number grouped closely together the detection is extremely difficult. All that the Sniper could do was to put a man on to keep the trees under constant observation. He chose Matthew Oldheim for the job because of the American's proved capacity for accurate observing, and gave him the best telescope of the sniping outfit to help him.

There was a clearness in the atmosphere on Oldheim's first day as watchdog, which boded the near approach of a rainy spell, but gave great assistance in observation work. The Sniper happened to be near the spot where the subaltern had been killed, when he heard a shot from Oldheim's loophole. He immediately fixed his periscope in position and looked towards the German line. Near the top of one of the trees under observation a German soldier was struggling. He was holding on to one of the branches and trying to swing himself behind the trunk. Another shot came from the loophole, and the German ceased struggling. He released his hold on the branch and crashed to the ground, falling across a tree which had been cut down by shell-fire. The crack of his breaking back whipped like a rifle-shot across no man's land.

Oldheim's success was hailed with great delight. No more dramatic snipe had taken place in the history of the battalion. One after another of his comrades slipped along to congratulate him, and so many of the officers took him away to their dugouts to give him a 'reviver', that the sergeant had to spirit the gallant fellow quietly down the communication trench and put him to bed. Everyone felt that it was the beginning of the end for the German section.[19]

The Canadian Corps, fighting on the Western Front, had a reputation of responding swiftly and decisively to German snipers. This paragraph by Ralph Hodder-Williams of Princess Patricia's Canadian Light Infantry, holding the line around St Eloi in January 1915, is indicative of the robustness of the Canadian response:

There was much sniping both by day and night, and in this the Germans had at first so great a superiority that Colonel Farquhar instructed Lieutenant W.G. Colquhoun, the Scout

Officer, to form a corps d'élite of marksmen to cope with the enemy's snipers. Lieutenant Colquhoun was given a roving command and the men whom he chose and organised under Corporal J.M. Christie became free lances who fought the German snipers with their own weapons. Their success was at once very marked – in a two-day tour near the Mound they accounted for seventeen of the enemy – and Lieutenant Colquhoun rendered good service in thus laying the foundations of a sniping section in which the Regiment always took great pride.[20]

By the autumn of 1915, Major Hesketh-Prichard was conscious of an increasingly wary attitude among German snipers, and accordingly, techniques were developed to bring them out into the open. During this period he was training officers of the 48th Division in the area around Hebuterne; this entailed long practical sessions in the front line. On one occasion he took a small group of students into the line, along with another sniper trainer and fellow big-game hunter, Captain G.M. Gathorne-Hardy (referred to as G.):

At the time of which I write the Germans were just beginning to be a little shy of our snipers on those fronts to which organization had penetrated, and it was clear that the time would arrive when careful Hans and conscientious Fritz would come very troglodytic, as indeed they did. We had, therefore, turned our minds to think out plans and ruses by which the enemy might be persuaded to give us a target. We had noticed the extraordinary instinct of the German officer to move to a flank, and thinking something might be made out of this, we collected all our officers and went back to the place where G. and I had spotted a Hun sniper or sentry at the end of a sap. A glance showed that he was still there.

I then explained my plan, which was that I should shoot at this sentry and in doing so, deliberately give away my position and rather act the tenderfoot, in the hope that some German officer would take a hand in the game and attempt to read me a lesson in tactics.

On either flank about 150 yards or so down the trench I placed the officers under instruction with telescopes and telescopic-sighted rifles, explaining to them that the enemy snipers would very possibly make an attempt to shoot at me from about opposite them. I then scattered a lot of dust in the loophole from which I intended to fire, and used a large .350 Mauser, which gave a good flash and smoke. As the sentry in the sap was showing an inch or two of his forehead as well as the peak of his cap, I had a very careful shot at him which G., who was spotting for me with the glass, said went about twelve inches too high.

The sentry, of course, disappeared, and I at once poured in the whole magazine at a loophole plate, making it ring again, and by the dust and smoke handsomely giving away my position. I waited a few minutes, and then commenced shooting again. Evidently my first essay had attracted attention, for two German snipers at once began firing at me from the opposite flank. At these two I fired back: they were almost exactly opposite the party under instruction, and it was clear that, if the party held their fire, the Germans would probably give fine targets. As a matter of fact, all that we hoped for actually happened, for the exasperated German snipers, thinking they had to deal only with a great fool, began to fire over the parapet, their operations being directed by an officer with an immense pair of field-glasses. At the psychological moment, my officers opened fire, the large field-glasses dropped on the wrong side of the parapet, as the officer was shot through the head, and the snipers, who had increased to five or six, disappeared with complete suddenness. Nor did the enemy fire another shot.[21]

Despite the general success of the British sniping recovery in 1915–16, German sharpshooters remained a threat; not all British battalions had good sniper sections at their disposal. Throughout the war, unwary British troops continued to be struck down by well-placed German bullets.

During the First World War, all armies used high-velocity bolt-action rifles as their standard infantry armament. Although there were minor differences, they were strikingly similar in design and performance. The German Mauser Gewehr 98 was a first-rate rifle which lent itself to sniping with the addition of telescopic sights. The British armed forces were issued with the SMLE (Short Magazine Lee Enfield) rifle (officially designated Rifle No. 1, Mk III), a reliable infantry weapon – much liked by the troops – but lacking in accuracy. As a result British snipers were usually issued with the Pattern '14 rifle (Rifle No.3, Mk I), which, although slower to load and fire, was more accurate. Snipers in the Canadian forces used the Ross rifle, a Canadian design not suited to general service use in the trenches, but whose superior accuracy made it suitable for sniping. Herbert McBride describes the issue of his rifle and sights:

> These Ross rifles were exceptionally accurate and dependable with the Mark VII ammunition we were then using. With the rifle I got a telescopic sight. Prismatic and mounted on the left side of the rifle, it was better than any other I had used. I also got a fine spotting 'scope and a tripod for the same. The spotting scope was pretty high magnification – about 36-power I believe. Each article – sight, telescope and tripod – had its separate sole-leather carrying case, with convenient straps for slinging them over the shoulder.[22]

Apart from superior quality rifles, snipers experimented with a variety of devices to improve performance or make sniping safer. Rifle rests comprised a rifle clamped to a fixed position in the trench parapet, its sights trained on a section of the enemy trench where nocturnal enemy activity was expected. During the night, the rifle would be fired at irregular intervals in the hope of hitting an enemy soldier who had otherwise considered himself safe in the darkness.

Periscopes provided safe observation over the trench parapet, and this principle was extended to rifle fire with the invention of the sniperscope (sometimes called the periscope rifle), an ingenious device for short-range sniping. Ion Idriess, an Australian soldier fighting at Gallipoli, describes the working of a sniperscope:

> I'm handling a periscope rifle now, it's the first time I've used one. The opposing trenches are so close that loopholes are useless to either side. Any loophole opened in daylight means an instant stream of bullets. So Jacko [a Turkish soldier] uses his periscope rifle and we reply with ours. A periscope [rifle] is an invention of ingenious simplicity, painstakingly thought out by man so that he can shoot the otherwise invisible fellow while remaining safely invisible himself. Attached to the rifle butt is a short framework in which two small looking glasses are inserted, one glass at such a height that it is looking above the sandbags while your head, as you peer into the lower glass, is a foot below the sandbags. The top glass reflects to the lower glass a view of the enemy trenches out over the top of the parapet. It is a cunning idea, simple and deadly, but I feel clumsy with it at first.[23]

The clumsiness experienced by Idriess would never be overcome. The sniperscope was not a precision system; accurate long-range shooting could only be carried out with a rifle in the shoulder of a trained sniper.

Loophole plates were bullet-proof rectangular steel plates - within which was cut a small hole sufficient to allow the sniper to aim his rifle — set into the trench parapet. Ideally, the plate was hidden in the parapet so that the enemy would be taken by surprise. If discovered, a loophole plate would produce caution in enemy troops and might encourage retaliation from mortars and artillery. McBride advised against plates facing directly towards the enemy, suggesting that they be sited at an angle to the line of the trenches: 'Enemy observers, using periscopes, are prone to scan the ground in their immediate front, depending on others to watch their own sectors. Thus, it is an easy matter to have the hole perfectly screened from the direct front, while giving a good field of fire along an adjacent portion of the line.'[24]

Duels between snipers using loophole plates were common on the Western Front. An example of this practice is provided by Frederick Sleath – the Sniper, aided by his sergeant and an experienced marksman called Old Dan, searches for an enemy sniper who has killed one of his men:

It was the Sniper who discovered the post sheltering the German responsible for the vacancy in his section. He spent most of the following morning in searching the German parapet opposite with minute care. There were many suspicious blotches of colour purposely made by the Germans to camouflage their posts, but the excellence of his prismatic periscope enabled him to decide on one particular point as the most likely to be the exterior of a loophole. Both the sergeant and Old Dan after a careful examination agreed with him, and it only remained to test the suspected spot.

The method for testing for a suspected loophole was quite simple. A shot fired against a steel plate gave a distinctive metallic clang, altogether different from the noise of a bullet striking a parapet or even a stone; and if the plate was there, no matter how skilfully camouflaged it might be, a bullet would prove its presence.

'Will you try a shot, sir?' said the sergeant, looking at the Sniper expectantly.

The invariable custom of the section was that the man who discovered an enemy sniping post should take the first shot at it. The Sniper at once took Old Dan's proffered rifle, and went into the nearest loophole.

'Look out, sir!' he heard the sergeant yell immediately after he had fired.

A long rectangular panel had opened in the German parapet. The daylight shone through the opening, and revealed the extent of the loophole quite plainly. He saw the long barrel of a German rifle appear in the gap, and after swinging about uncertainly from one end of the sector to the other, a little puff of uncombusted cordite came from the muzzle and a shot rang on his shield within an inch of the loophole.

'Two inches low left,' called the sergeant, as the Sniper fired again.

The Sniper corrected his aim. His finger was pressing full on the trigger when a second shot from the German struck on the loophole. The shock made him involuntarily complete the pull.

Somehow or other the Sniper had retained his aim in spite of the accidental pressing of the trigger, and the bullet had gone home. He could see a large white splinter from the wooden casing of the rifle barrel, lying below the loophole, and the rifle itself was cocked up in the air by the weight of the butt and slowly slithering out of sight into the trench.

It was his first snipe. He had brought it off at considerable risk to himself. They were proud of the pluck and skill which their officer had displayed. But this taking of a man's life shook him considerably. A pull at his flask cured him, however, and he was as busy as ever supervising the activities of his men.[25]

This account reveals the inexperience of the German sniper in casually opening his loophole and revealing himself to the British opposite because of the light shining behind him. The standard practice, as taught by a Hesketh-Prichard or Crum, was to 'gag' the loophole when not in use, and place a curtain behind the position so that when the loophole was opened no light would shine through. It was an elementary precaution, but one which saved lives.

The correct siting of sniper posts could also make the

difference between life and death. Once the enemy had discovered a troublesome post, retribution would inevitably follow, if not by another sniper, then by an artillery shoot which would blow the position to smithereens. Front-line posts had constantly to be changed, and a proficient sniper would move from a variety of shooting positions on a regular basis.

Many top snipers preferred to work slightly behind the lines, adopting positions that gave them a better field of view than that found in a front-line trench. Firing from trees and buildings was generally discouraged as these positions were too obvious, and too clearly visible to the enemy. If possible, well-camouflaged sites in the open were to be chosen. This could often mean extensive shovelling work from the snipers involved. After one of their sniper posts had been discovered, McBride and his observer partner moved to another prepared position:

> This one was right out in an open field but was reached by a tunnel-like trench which we had dug during many nights of hard labour. The tunnel was about three feet deep and nearly as wide, and extended for more than one hundred feet out beyond the corner of the last of the outlying buildings. In digging it, we would first take off the sod and lay it aside, then excavate the earth to the desired depth, carrying all the dirt back and hiding it in the building. Then we covered the top with pieces of board from the wrecked building, put on a little earth and replaced the sod. At the end, where we had our 'nest', was a good-sized chamber, big enough for the two of us to stretch out comfortably, and with two holes, one for the telescope and the other for the rifle. Small bushes and clumps of grass immediately in front (which were taken into account when planning it) effectively served to screen it from enemy observation.[26]

Most snipers worked in pairs, one shooting, the other acting as an observer, picking out targets and correcting the fall of shot. The recoil of the rifle and attendant muzzle blast made it difficult for

the sniper to observe the effect of his own shot, and consequently
he relied on his partner (equipped with a high-powered telescope)
to inform him of success. Hesketh-Prichard made these comments
on the ability of an observer to tell whether the target had been
hit:

> Continued observation showed that a man shot in ordinary
> trench warfare very rarely either threw up his hands or fell
> back. He nearly always fell forward and slipped down. For
> this the old Greek rendering is best, 'And his knees were
> loosened'.
>
> We soon found that a very skilled man with a telescope
> could tell pretty accurately whether a man fired at had been
> hit, or had merely ducked, and this was the case even when
> only the 'head of the target' was visible. The idea of how to
> spot whether a German was hit or not was suggested by big-
> game shooting experiences. An animal which is fired at and
> missed always stands tense for a fraction of a second before it
> bounds away, but when an animal is struck by the bullet
> there is no pause. It bounds away at once on the impact, or
> falls.
>
> In dealing with trench warfare sniping, a very capable
> observer soon learned to distinguish a hit from a miss, but
> there were naturally many observers who never reached the
> necessary degree of skill. A reason once advanced for claiming
> a hit was that the Germans had been shouting for stretcher-
> bearers, but a question as to what was the German word for
> stretcher-bearer brought confusion upon the young sniper,
> whose talents were promptly used elsewhere![27]

As part of the observational skills that were second nature to a
sniper, Herbert McBride prepared these notes on the sound and
nature of rifle fire:

> The effect of any bullets fired from the German Mauser was
> very similar to that of the 150-grain bullet fired from the

[US] Springfield. At short distances, due to high velocity, it does have an explosive effect and not only that effect but, when it strikes, it sounds like an explosion. Bullets may be cracking viciously all around you when, all of a sudden, you hear a 'whop' and the man alongside you goes down. If it is in daylight and you are looking that way, you may see a little tuft of cloth sticking out from his clothes. Wherever the bullet comes out, it carries a little of the clothing – just a fuzz – but it is unmistakable.[28]

Although the majority of sniper action took place along the Western Front, there were other theatres of war where sniping was prominent. One such area of operations was Gallipoli, a narrow peninsula in European Turkey where British, Commonwealth and French forces were pitted against the defending Turks. The Allies, short of artillery and machine-guns, were consequently more reliant than usual on rifle fire.

Among the Turkish Army were many fine shots, and operating from well-prepared defensive positions they usually had the edge in the sniper war. The Turks were helped by the carelessness of Allied troops who regularly exposed themselves to the sniper's bullet. Only slowly, through the adoption of greater caution in the trenches and the adoption of a counter-sniper policy, did the Allies come to terms with the Turkish snipers. None the less, the Turks remained determined opponents as this extract from Thomas Baker demonstrates. Baker was a private in the Chatham battalion of the Royal Navy Division, and he describes the problem of Turkish 'stay-behind' snipers around Shrapnel Valley:

We hadn't managed to clean up all the snipers. There were still snipers in the bush behind the front line. They used to pick quite a number of men off before they got wiped out, because they couldn't get out once they were behind the front line and we picked them off gradually. In the dark you used to watch for the flash of their rifles; that betrayed their positions to us.[29]

The Australians proved particularly adept at sniping, their foremost marksman being Billy Sing, a trooper in the 5th Light Horse who amassed a score of over 150 kills while at Gallipoli. Ion Idriess, a sniper from the 5th Light Horse, acted as a spotter for Sing on some occasions:

> He is a little chap, very dark with a jet-black moustache and a goatee beard. A picturesque-looking man-killer. He is the crack sniper of the Anzacs. His tiny possy [position] is perched in a commanding position high up in the trench. He does nothing but sniping. He has a splendid telescope and through it I peered across at a distant loophole, just in time to see a Turkish face framed behind the loophole. He disappeared. Father along the line, I spotted a man's face framed inquiringly in a loophole. He stayed there. Billy fired. The Turk vanished instantly, but with the telescope I could partly see the motion of men inside the door picking him up. So it was one more man for Billy's tally.[30]

Private Baker relates an instance of Australian bravado in winkling out a Turkish sniper:

> There was a Turkish sniper who was picking up men, man after man. To deal with this an Australian ran along the trench, with a dummy on his back, just when it was beginning to get dark, to attract the sniper's fire. We were watching to find out where the sniper was firing from – you could see the flash. We found he was up a tree, and all of us who could be relied upon to hit something fired into this tree. He was dressed in green but we were able to shoot him. The Aussie who attracted this fire was the bravest I've ever seen, but he got away with it. I wouldn't have done it for all the tea in China.[31]

Following the withdrawal of Allied troops from Gallipoli in 1916, the focus of the war against Turkey moved to the Middle East. In

1917, British and Commonwealth forces began a major offensive into Palestine, a campaign which eventually saw the defeat of the Turks and the capture of Jerusalem. Ion Idriess and his regiment of the 5th Light Horse were part of this offensive. In an extended narrative of a counter-sniper operation, Idriess describes the patience and talent for observation required in stalking a first-rate opponent, and the feelings that run through a sniper's mind as he closes with his quarry:

I was back in the galloping squadrons as we swung under cover behind some low hilly rises during the Gaza-Beersheba stunt. Our squadron hastily dismounted at the extreme end of one of these. A man immediately collapsed with a bullet through the lung. There was a rush to pull the panting horses as close under the sheltering rise as possible, for the dread thought of 'sniper' flashed through every mind. The enemy were lining the low hills in front of us, but our man had been shot from out on the right flank.

'Crack! Thud!' A horse crashed on its belly as a high-velocity bullet whistled into its flank. A good-looking chestnut reared surprisingly in the air, pawed frantically in an attempt to regain its balance, then crashed backward to earth, the soft brown eyes growing pitifully big as it struggled gamely to rise. The trooper-owner stared down at it, intense feeling spreading over his hard, drawn face. A mate rushed out from cover and jerked the man's arm; even as he did so a vicious, crackling whistle had come and gone. The emu feathers on the trooper's hat fell to his shoulder, sheared clean off to the hat rim.

Very gingerly I looked from the side of the rise through a pair of splendid field-glasses, 'souvenired' from an Austrian officer of artillery. I examined the field of barley on our right flank, whence apparently, the shots had come. It was a beautiful, sun-kissed field. Emerald green. The crop just about a foot high, and as level as a billiard table. High in the clear air above the centre of the little field a brown lark

trilled as only larks can. But of life in the field there was not
a sign. I well knew that a trained man can lie perfectly still
facing you not a hundred yards away on hard, brown earth
without even a grass blade on it, and you can stare at the
muzzle of his rifle for ten minutes without seeing him.

Between us and the edge of the field, just at the foot of
the gentle slope which ran down from our sheltering rise,
and only a hundred yards away, was a narrow wadi – or gully
as we would call it in our own good language. Throwing off
haversack and water bottle, I suddenly jumped from behind
the rise and tore down the slope in a crouching run, jerking
from left to right, not for three seconds keeping to one
straight line. The crackling hisses as the sniper tried to get
me by rapid fire were reminiscent of the breath of a white-
hot iron. I landed sprawling in the sheltering wadi-bottom,
and gasped in long, thankful breaths.

Pulling barley from the edge of the bank was the next
thing; then arranging the foot-long stalks in a row around
the hatband, and carefully spreading barley in flat sheaves
over my back under and over the bandolier and bayonet-belt.
Then came the stealthy climb up the wadi-bank and into the
barley field. First the length of the rifle (how heavy the
familiar weapon quickly became!) poked gently on ahead,
full-length of the arms, through the barley-stalks; then rested
gently on the earth, palms of the hands cupped protectingly
around the firing mechanism. Then careful craning forward
of the head, chin pressed to the ground; then dragging of the
entire taut body along by leverage of the elbows. Just one
foot advance for each drag, just one inch at a time; chin,
chest, belly and toes pressed tight against the earth, while
the heart thumped.

My mates back on the rise would go through all the old
tricks to attract the waiting sniper's fire and attention: a
gently moving hat held just above the skyline with a stick; a
rifle poked above a sand mound with a hat slanting on the
butt. If this had no effect, a man would dash from cover,

then back again, lively — anything so that I might locate the sniper by the crack of his rifle.

Without lifting chin from the ground, my eyes would naturally rise slightly to the sky on each pause for breath. Each time they would alight on the carolling lark, and it seemed as I wormed further and further into the field to hover constantly over one particular spot. I watched the lark for what seemed a long time; then a breathless realization gripped the mind and tingled through the body until it tautened at the roots of the scalp. I felt my mouth open a little and eyes suddenly widen as I pressed back the rifle safety-catch and gripped hard the splendid weapon. I wormed a little further on. Slow, tense minutes passed in withdrawing the field-glasses and carefully raising them above the ground.

Then something moved. It was only the turn of his cheekbone, but it allowed me to focus right into the eyes of my man. Only partly could I see the big brown nose, hawk-shaped, for two twisted barley-stalks camouflaged his black burnous. The perfectly shaped tiny black dot of his rifle-muzzle I could see, and below the telescopic sight the bony brown finger-knuckles that gripped around the weapon. A Bedouin, with the eyes of a hawk!

Then the calculation on which one life would depend. Dearly I would have loved to level the rifle foresight fair between those two black eyes — the desire grew almost overwhelming. But many barley-stalks besides the two crossed ones were in the way. Such a tiny thing might deflect a bullet — and a man would be allowed only one shot. I could see the butt of the two stalks as they crossed down his nose in a straight line past the centre of his chin where the black beard hid them.

'Aim right at the butt of the stalks,' whispered the mind, 'exactly where the beard covers them. The bullet then should strike that little hollow below the throat at the base of the neck between the two bones, and go right down through his body. He should never move after that!'

Then the indrawing of a deep breath, the raising of the rifle, the easing of the racing nerves as the familiar weapon settled its iron-shod butt reassuringly into the hollow of the shoulder, the absolute steadiness as the trigger-finger took the 'first pull' and the foresight lowered on the barley-stalk down past the eyes, past the mouth, slowly past the chin, until, engaging in the rear sight, stopped dead where the beard hit the butt of the barley-stalk.

'Crack!'

I bounded out of the barley and was on the spot even as he rolled over. He was dying. His flashing black eyes fastened on mine in a gaze of instant realization and deathless hate. He attempted to raise his arm, the sinews tautened in the thick brown wrist as he tried vainly to clench his fist. I knew – he was beseeching all the curses of the Prophet upon this Christian dog who had taken his life. Looking down at him, like a great fallen hawk in the crushed barley, I felt no remorse; only hot pride that in fair warfare I had taken the life of a strong man.

Quickly I looked him over for the inevitable souvenirs. Strung on a camel sinew around his neck were thirty-eight identification disks, mostly those of British troops, but with a sprinkling of Australian and one Maoriland badge. A Mohammedan goes to Paradise if he can kill a Christian. So this good religionist had thirty-eight keys to the pearly gates. There was a special silver medal also, and a parchment deed of recognition from the Sultan.

Presently I turned and examined the barley-roots closely. But it was not until a couple of fellows from the squadron came running over that we found what we sought. Within a foot of the Bedouin's body, cunningly interwoven between four stalks of barley, was a little nest, and in it one solitary fledgling, its eyes still shut, but hungry mouth wide open. Such an insignificant thing to cause the death of a man![32]

Accounts such as this may imply that all sniping operations were conducted with the greatest ferocity. Although none can doubt the scale and intensity of the First World War, there were often moments when the fighting slackened and the ordinary soldier adopted a non-belligerent policy towards the enemy. The Christmas Truce of December 1914 was the best-known instance of this tendency, although there were many other minor examples. Snipers, by the nature of their occupation, minimized the 'live-and-let-live' approach to trench warfare. This was not always the case, however. The history of a battalion in the 48th Division reveals how its colonel, visiting damaged front-line trenches, found to his disgust that:

> Not a shot was being fired at the enemy as they dodged across the gaps in their trench. They in their turn, were allowing our men to cross the gaps in our trench without molestation. This was too much for the colonel. The first sentry he came across happened to be the sniper now returned to duty with his company. He received the full blast of the colonel's indignation: 'Man alive, what are you there for? Don't you see these Huns?' To which the sentry diffidently replied: 'I ain't a sniper now, sir.'[33]

At times the natural humour of the troops broke through, helping relieve the monotony of trench life. From his experiences at Gallipoli, Ion Idriess recalled this incident:

> We have been chuckling over a bit of fun away at Quinn's post. The boys rigged up quite an inviting bull's eye and waved it above the trench. Each time the Turks got a bull, the boys would mark a bull. For an outer the boys marked an outer, for a miss they yelled derision. The Turks laughed loudly and blazed away like sports. After a while an officer came along and of course had to be a spoil sport.[34]

Another similar incident of 'target-shooting' humour was recorded
in July 1915 by Sergeant-Major Hanlon of the 7th Battalion of
the Royal Sussex Regiment, although the storyteller only just
survived to tell the tale:

> There were a number of periscopes on top of the parapet of
> the German trenches, including one very large one, obviously
> a dummy, just in front of the brewery at Frélinghien. This
> particularly attracted me, so one afternoon I went round with
> a rifle fitted with a sniperscope and had a shot at it.
> Immediately after firing the Germans turned it sideways and
> waved a flag, signalizing a washout. I fired another round
> and a large broom was waved to and fro and then lowered. I
> was firing from the left of Lumbago Walk (so called because
> of the struts across it, which necessitated one's bending
> almost double for some yards) when suddenly a shot from
> the crater on my right front went straight through the top
> sandbag not far from my head.[35]

Incidents such as these were not confined to the First World War.
Others were recorded in conflicts as far apart as the American
Civil War and Vietnam. They were simply the responses of bored
soldiers, facing an enemy in a static position for weeks or months
on end. The tendency of the human spirit to get the better of
military prudence is a common phenomenon of warfare – even
when enemy snipers are active. One example is noted by William
March, whose novel recounted the stories of the soldiers from
Company K, part of an American battalion fighting on the
Western Front in 1918. The story told by Private Leo Hastings is
a good example of such recklessness:

> All that morning the German sniper shot at me. I would
> stick my head up, or walk across the open space, and there
> would come a faint ping and a bullet would pass harmlessly
> over my head. Then I would stop in my tracks and stand
> there a full two seconds, or suddenly take a step backwards

and a step to the side. I would walk that way up and down the parapet of the trench, laughing at the sniper. I knew I had him so sore at me that he was almost ready to break down and cry. I'd shot with telescopic sights myself and I knew no sniper in the world could hit a man who varied his stride as I did, unless the sniper could figure out in advance the man's system, and he's got as much chance of doing that as he has of breaking the bank at Monte Carlo.

'I'll stand here and let him take pot shots at me all day,' I said, 'he can't hit me in a thousand years.' See, as I stand now, he's got me covered. But wait! He's got to figure his distance, taking an angle between that dead tree and the farmhouse, probably. Now he's got it all doped out. He's taking his windage and calculating elevation. Now he's all ready to plug me, but by stepping one pace to the right, or doing this clog step I upset all his calculations. 'See,' I said 'there goes his bullet, two feet to the right. He can't hit me to save his neck.'[36]

Despite the constant threat of death from the enemy, some soldiers bridled at shooting an unsuspecting human target; it was unnerving to look down a telescopic sight and see another human being going about his business, knowing that he would probably be dead a few seconds later. Other men, however, had few concerns with the morality of sniping, comparing it to another form of hunting. Herbert McBride summed up such a view: 'I assure you that when I was behind the rifle, the principle feeling was one of satisfaction and excitement of the same kind that the hunter always knows. That's the spirit. That's what makes good riflemen and good soldiers.'[37]

As an élite infantryman, the sniper may have been respected by his fellow soldiers but he was rarely liked. In part, this was a consequence of the natural revulsion towards killing an unsuspecting man; soldiers in the front line tended to share an affinity with the enemy, both of whom were fellow sufferers in the misery of trench warfare. More important, however, the sniper brought retaliation from the enemy. An instance of the uneasy relationship between sniper and ordinary soldier is provided by F.A.J. Taylor, engaged in observation duties shortly after his arrival in the front line:

Our lads in that front-line trench were never happy at our presence. They were quite naturally keen to maintain the status quo and thought we might stir up an enemy strafe and were apprehensive, venting their feelings by the most blood-curdling curses in our direction. Their fears were not unfounded as one experience showed. I was gaining experience in using the telescope when one day I became suspicious of a small, dark triangular patch in a mound of earth about three hundred yards in front. Staring at it for some time I couldn't make out what was a light circular patch in the middle of it. When to my utter amazement I saw very clearly the side of the face of a German, with light brown hair. His profile showed as he turned away. I kept on watching as I speculated, then suddenly I saw a wisp of smoke and instinctively ducked as his bullet zipped viciously through the sandbags over my head.

Then I realized that I had been gazing at an enemy telescope in their observation post while at the same time he had been watching me. Just then our bespectacled Intelligence Officer made one of his infrequent inspections of our post and I told him what had happened with great excitement. He asked for my loaded rifle which he poked over the parapet as he mounted the fire step. But before he was ready, another vicious zip came and a bullet went past his ear flicking dust in his face. He promptly dropped back into the trench and looking very white handed me back my trusty rifle. Two Tommies nearby uttered loud obscenities and uncomplimentary remarks. Our officer hurried away down the trench to make his report to the gunners' officer at Battalion HQ. We had not long to wait before the mound was heavily shelled and went up in smoke. It was without doubt a small enemy redoubt with a forward observation post, quite some distance in front of the enemy front trench.

We watched as we saw a few surviving grey-clad figures hurry away and go to ground, and that was the last we saw of that enemy observer. My turn of duty over I moved away and as I passed the two lads in the adjoining bay, one said, 'It's all right for you buggers, but you go away and leave us to get all the shit.' And it came true as Jerry plastered that bit of trench in revenge.[38]

Another British sniper, William Carson-Catron, observed that as snipers 'moved about their own front line they would be told, "Push-off now, do your stuff somewhere else".' Comments such as this were typical, but to highly motivated men, they were easily shrugged off. Snipers were far more fearful of capture by the enemy, knowing they could expect no mercy. One Australian infantrymen at Gallipoli routinely recorded the capture of a Turkish sniper 'who was shot without trial'. Captain J.C. Dunn, the medical officer of a battalion of Royal Welch Fusiliers, collated accounts of the war from fellow battalion members, and recorded this description of the fate of a brave but ultimately negligent German sniper captured in October 1914:

A lot of sniping from houses had to be put up with, so had the shelling, for our guns had not enough ammunition to cope with them. But the circumstances in which some men had been shot led to the belief that a sniper was behind us, so Stanway, who had several snipers to his credit, took a few men to make a search. While they were halted beside a large straw rick one of the men noticed some empty German cartridge cases at his feet. On thrusting their bayonets into the rick the party was rewarded by a yell and a German coming out headlong. Inside was a comfortable hide, having openings cleverly blocked with straw, and a week's supply of food. The sniper could come out at night for exercise and water. Only his carelessness, with his used cartridges, cost him his life, for he was finished there and then.[39]

Snipers exploits were, in official terms, largely unrecorded and unrewarded. Operating on their own, snipers conducted most of their work out of sight of officers, who had the responsibility of making awards. The only Victoria Cross awarded to a sniper during the First World War went posthumously to Private Thomas Barratt of the 7th Battalion of the South Staffordshire Regiment. Barratt was killed in action just north of Ypres on 27 July 1917; the *London Gazette* printed this sparse citation:

> For most conspicuous bravery when as a scout to a patrol he worked his way towards the enemy's line with the greatest gallantry and determination, in spite of continuous fire from hostile snipers at close range. These snipers he stalked and killed. Later his patrol was similarly held up, and again he disposed of the snipers.
>
> When during the subsequent withdrawal of the patrol it was observed that a party of the enemy were endeavouring to outflank them, Private Barratt at once volunteered to cover the retirement, and this he succeeded in accomplishing. His accurate shooting caused many casualties to the enemy, and prevented their advance.
>
> Throughout the enterprise he was under heavy machine-gun and rifle fire, and his splendid example of coolness and daring was beyond all praise. After safely regaining our lines, this very gallant soldier was killed by a shell.[40]

In any war, close observation of the enemy was essential, to discover his strengths and dispositions and guess at his intentions. Snipers were particularly useful in this role, and as the war progressed the reconnaissance element of their work increased. Equipped with 6-power binoculars and 30-power telescopes, snipers were able to gain information impossible to discern with the naked eye. Equally important was the quality of such observation. An untutored man could survey a scene for hours with the best telescope and still be oblivious to what a good sniper could see straight away. An instance of the superiority of the trained observer is provided by F.A.J. 'Tanky' Taylor, who had been sent to an observation post in the Ypres salient by his experienced sniper sergeant, Howard 'Pincher' Hill:

> Out front we could see the two opposing front lines and our job was to observe and note any troop movements, new construction, and all enemy shell fire. So here I was in 1918, for the first time, in the observation post in the ramparts of Ypres, about halfway between the Menin and Lille gates. My job was to scrutinize the landscape and make my report to our sergeant. I was kept pretty busy noting the falling and exploding shell, and in between I gazed out at the row of small hills, rising from the Flanders' plain.
>
> Gazing intently, with both eyes open in the correct manner, an hour passed and I saw no movement or any Germans. When Pincher appeared to see how I was getting along, and asked me if I'd seen anything, I replied, 'Not a bloody thing.'
>
> He took over the telescope and started to slowly scour the landscape. He said, 'Nothing eh, so youm be seeing nothing, ah um.' Then he turned to me and said, 'The place is lousy with 'em, take a good look, Tanky.' He told me where to look and the objects which helped. Sure enough I could see green-grey-clad figures moving about. I could now see Germans all over the place. He had the skill, ability and experience to spot the most likely place, and the movement which was always the giveaway.[41]

The most able observers were the Lovat Scouts, raised in the Highlands of Scotland from gamekeepers trained in stalking deer.

Hesketh-Prichard was a great admirer of the Lovat Scouts, and worked with them extensively. One scout, Corporal Donald Cameron, was particularly adept:

> He was a very skilful glassman, and as such was of continual assistance to me. I remember one day when we were trying some aspirant reinforcements for Lovat Scout Sharpshooters and were looking through our glasses at some troops in blue uniforms about six thousand yards away, most of the observers reported them as 'troops in blue uniforms' but Cameron pointed out that they were Portuguese. His reasoning was simple. 'They must be either Portuguese or French [who wore blue uniforms],' said he, 'and as they are wearing the British steel helmet, they must be Portuguese.'[42]

Hesketh-Prichard provided a further example of the application of informed intelligence to simple observation:

> The most brilliant piece of deduction that I came across was that of an officer of the Royal Warwickshire Regiment, and it had a remarkable sequel. At one point of a supposed disused trench, a cat was observed sunning itself upon the parados. This was duly reported by the observant sniper, and in his log book for three or four days running came a note of this tortoise-shell cat sunning itself, always at the same spot.
>
> The Intelligence and Sniping Officer of the battalion, on reading his entries, made his deduction, to wit, that the cat probably lived nearby. Now at that part of the British line there was a terrible plague of rats, which was probably at least as troublesome upon the German side. So our officer deduced that the cat was a luxury, and this being so, it had most certainly been commandeered or annexed by enemy officers and probably lived in some enemy officer's headquarters – possibly a company commander's dugout.
>
> Some aeroplane photographs were taken and studied, with a result that an enemy headquarters was discovered,

located and duly dealt with by one of the batteries of howitzers which make a speciality of such shoots.[43]

At the war's end, sniping on the Western Front had developed into an advanced military craft. The sniper was capable of accurate shooting at long range; he was also a master of fieldcraft, a capable cross-country navigator and a skilled observer. Before 1914 the sniper had been an inspired amateur, but the war transformed his calling into one of the many military disciplines that were to beome a part of any modern army.

Chapter Five

The Second World War

Sniping in the Second World War was remarkable for its geographical diversity. Snipers operated across the globe, from the hedgerows of Normandy to the ruins of Stalingrad, and from Italy's Apennine mountains to the jungles of New Guinea.

The earlier stages of the war were dominated by the fast-moving tactics of Blitzkrieg, where sniping was less relevant. But by 1942 snipers were beginning to re-emerge as a potent force on the battlefield. The Canadian/British raid on the French coastal town of Dieppe in August 1942 was a disaster for the Allies: poor training and inadequate equipment prevented the main body of troops from getting off the beach, and casualties were high. Both sides used snipers to good effect, however. The war correspondent Wallace Reyburn commented: 'I talked to a half-caste Indian named Huppe. He'd spent most of his life in the backwoods of Canada as a hunter and trapper and he'd become an outstanding rifle-shot. Just before I arrived on the scene he'd picked off ten Germans in the space of half an hour or so.'[1]

On the flanks of the main force, units of Commandos had been assigned to destroy coastal batteries and other German defences. No. 4 Commando, on the right flank, was particularly successful in its mission, deploying its snipers effectively. Derek Mills-Roberts was a senior officer in 4 Commando, and his unit's contribution to this local triumph was described by the historian Ronald Atkin:

As Mills-Roberts and his men broke clear of the woods they dropped to the ground. Ahead of them was the perimeter wire and behind that, at a distance of 170 yards, the huge barrels of the battery. To Mills-Roberts's right, on the edge of the wood, stood a two-storey barn. Taking two snipers with him, he worked his way round to the barn and, from the upper storey, was presented with a clear view of the six guns and crews working them.

The first sniper's bullet was to be a signal for all-out fire to be opened on the Germans and Mills-Roberts gave the order to proceed: 'The marksman settled himself on a table, taking careful aim. These Bisley chaps are not to be hurried. At last the rifle cracked. It was a bullseye and one of the Master Race took a toss into the gun pit. His comrades looked shocked and surprised – I could see it through my glasses. It seemed rather like shooting one of the members of a church congregation from the organ loft.'

The Germans reacted quickly. A flak tower on stilts sprayed the wood with tracer, and mortars caused casualties among the Commandos around the perimeter wire. The flak tower was silenced by an anti-tank rifle and soon the raiders brought into action their own 2-in mortar, operated by Troop Sgt-Major Jimmy Dunning with spectacular results. The third shot struck a stack of cordite which ignited with a mighty explosion.

Gleefully, the Commandos sniped at the battery crews frantically trying to extinguish the fires started by the explosion. The most daring was L/Cpl Dick Mann who, his hands and face painted green, crawled forward over open ground with a telescopic rifle and sniped at the crews from a fully exposed position.[2]

A small number of US Rangers had accompanied the Commandos to gain experience of combined operations. Among these was Corporal Franklin Koons, who had taken up a position in the barn: 'I found a splendid spot for sniping, just over the manger,

and I fired through a slit in the brick wall. I fired quite a number
of rounds on stray Jerries and I am pretty sure I got one of them.'³
As a result, Corporal Koons was credited with being the first US
soldier to kill a German in the Second World War.

While the fighting around Dieppe was taking place, German
panzer columns were closing on the Russian city of Stalingrad. By
September 1942 they were fighting the Red Army within the city
boundaries in one of the most titanic and crucial battles of the
war. Aerial and artillery bombardments turned Stalingrad into
rubble, a terrain where sniping prospered.

Both sides deployed crack shots in large numbers and many
became experts in street sniping. The most famous of the Russian
snipers was Vasili Zaitsev, a hunter from the Ural mountains who
was credited with killing 242 Germans in Stalingrad. A number
of accounts have been written of Zaitsev's exploits, but perhaps
the best comes from General V.I. Chuikov, who commanded the
beleaguered Soviet garrison at Stalingrad:

> We paid particular attention to the development of a snipers'
> movement among our troops. The Army Military Council
> supported this move. The Army's newspaper *In Our Country's
> Defence* published daily figures of the number of the enemy
> killed by our snipers, and published photographs of outstand-
> ingly accurate marksmen.
>
> I met many of the well-known snipers, like Vasili
> Zaitsev, Anatoli Chekhov and Viktor Medvedev; I talked to
> them, helped them as far as I could and frequently consulted
> them. These well-known soldiers were not distinguished in
> any particular way from the others. Quite the reverse. When
> I first met Zaitsev and Medvedev, I was struck by their
> modesty, the leisurely way they moved, their particular
> placid temperament, the attentive way they looked at things;
> they could look at the same object for a long time without
> blinking. They had strong hands: when they shook hands
> with you they had a grip like a vice.
>
> The snipers went out 'hunting' early in the morning to

previously selected and prepared places, carefully camouflaged themselves and waited patiently for targets to appear.
They knew that the slightest negligence or haste would lead
to certain death: the enemy kept a careful watch for our
snipers. They used very few bullets, but every shot from a
sniper meant death or a wound for any German caught in his
sights.

Vasili Zaitsev was wounded in the eyes. A German sniper
obviously took a lot of pain to track down the Russian
'hunter' who had about three hundred German deaths to his
credit. But Zaitsev continued to be an enthusiast of 'sniper-
ism'. When he came back to active service after his injury he
went on selecting and training snipers, his 'young hares'.

The activities of our snipers caused the German generals
a lot of disquiet, and they decided to turn this military craft
against us as well. This happened at the end of September.
One night our scouts brought in an identification prisoner,
who told us that the head of the German school of snipers,
Major Konings, had been flown in from Berlin and given the
task, primarily, of killing the leading Soviet sniper.

'The arrival of the Nazi sniper', says Vasili Zaitsev, 'set
us a new task: we had to find him, study his habits and
methods, and patiently await the right moment for one, and
only one, well-aimed shot.

'In our dug-out at nights we had furious arguments
about the forthcoming duel. Every sniper put forward his
speculations and guesses arising from his day's observation of
the enemy's forward positions. All sorts of different proposals
and "baits" were discussed. But the art of the sniper is
distinguished by the fact that whatever experience a lot of
people may have, the outcome of an engagement is decided
by one sniper. He meets the enemy face to face, and every
time he has to create, to invent, to operate differently.

'Just the same, where was the sniper from Berlin? – we
asked ourselves. I knew the style of the Nazi snipers by their
fire and camouflage and without any difficulty could tell the

experienced snipers from the novices, the cowards from the stubborn, determined enemies. But the character of the head of the school was still a mystery for me. Our day-by-day observations told us nothing definite. It was difficult to decide on which sector he was operating. He presumably altered his position frequently and was looking for me as carefully as I was for him. Then something happened. My friend Morozov was killed, and Sheykin was wounded, by a rifle with telescopic sights. Morozov and Sheykin were considered experienced snipers; they had often emerged victorious from the most difficult skirmishes with the enemy.

'Now there was no doubt. They had come up against the Nazi "super-sniper" I was looking for. At dawn I went out with Nikolay Kulikov to the same positions as our comrades had occupied the previous day. Inspecting the enemy's forward positions, which we had spent many days studying and knew well, I found nothing new. The day was drawing to a close. Then above a German entrenchment unexpectedly appeared a helmet, moving slowly along a trench. Should I shoot? No! It was a trick: the helmet somehow or other moved unevenly and was presumably being held up by someone helping the sniper, while he waited for me to fire.

'"Where can he be hiding?" asked Kulikov, when we left the ambush under cover of darkness. By the patience which the enemy had shown during the day I guessed that the sniper from Berlin was here. Special vigilance was needed.

'Nikolay Kulikov, a true comrade, was also fascinated by this duel. He had no doubt that the enemy was there in front of us, and he was anxious that we should succeed. On the third day, the political instructor, Danilov, also came with us to the ambush. The day dawned as usual: the light increased and minute by minute the enemy's positions could be distinguished more clearly. Battle started close by, shells hissed over us, but, glued to our telescopic sights, we kept our eyes on what was happening ahead of us.

'"There he is! I'll point him out to you!" the political

instructor suddenly said, excitedly. He barely, literally for one second, but carelessly, raised himself above the parapet, but that was enough for the German to hit and wound him. That sort of firing, of course, could only come from an experienced sniper.

'For a long time I examined the enemy positions, but could not detect his hiding-place. From the speed with which he had fired I came to the conclusion that the sniper was somewhere directly ahead of us. I continued to watch. To the left was a tank, out of action, and on the right was a pill-box. Where was he? In the tank? No, an experienced sniper would not take up position there. In the pill-box, perhaps? Not there either – the embrasure was closed. Between the tank and the pill-box, on a stretch of level ground, lay a sheet of iron and a small pile of broken bricks. It had been lying there a long time and we had grown accustomed to it being there. I put myself in the enemy's position and thought – where better for a sniper? One had only to make a firing slit under the sheet of metal, and then creep up to it during the night.

'Yes, he was certainly there, under the sheet of metal in no man's land. I thought I would make sure. I put a mitten on the end of a small plank and raised it. The Nazi fell for it. I carefully let the plank down in the same position as I had raised it and examined the bullet hole. It had gone straight through from the front; that meant that the Nazi was under the sheet of metal.

'"There's our viper!" came the quiet voice of Nikolay Kulikov from his hide-out next to mine.

'Now came the question of luring even a part of his head into my sights. It was useless trying to do this straight away. Time was needed. But I had been able to study the German's temperament. He was not going to leave the successful position he had found. We were therefore going to have to change our position.

'We worked by night. We were in position by dawn.

The Germans were firing on the Volga ferries. It grew light quickly and with daybreak the battle developed with new intensity. But neither the rumble of guns nor the bursting of shells and bombs nor anything else could distract us from the job in hand.

'The sun rose. Kulikov took a blind shot: we had to rouse the sniper's curiosity. We had decided to spend the morning waiting, as we might have been given away by the sun on our telescopic sights. After lunch our rifles were in the shade and the sun was shining directly on to the German's position. At the edge of the sheet of metal something was glittering: an odd bit of glass or telescopic sights? Kulikov carefully, as only the most experienced can do, began to raise his helmet. The German fired. For a fraction of a second Kulikov rose and screamed. The German believed that he had finally got the Soviet sniper he had been hunting for four days, and half raised his head from beneath the sheet of metal. That was what I had been banking on. I took careful aim. The German's head fell back, and the telescopic sights of his rifle lay motionless, glistening in the sun, until night fell.'[4]

Erich Kern was a witness to German sniper operations on the Narva bridgehead near Leningrad. The account reflects the grimness of the fighting on the Eastern Front, which in this instance bears a more than passing resemblance to conditions on the Western Front:

We were lying hard up against the enemy, in places no more than forty yards from their lines. It was a sniper's war. Dead silence reigned in the narrow strip of shell-torn land between us and the Soviet sap. Occasionally a heavy shell rumbled across the lines, and occasionally we heard the swish of the 'Hitler Scythe,' as the Russians called our machine-guns.

Then silence, heavy and oppressive, dropped again over the spectral landscape. Rain fell in a steady monotony out of a pale grey sky. Water stood knee-high in the trenches and gurgled over the edge of our gumboots.

Undeterred by the weather, men stood here and there in the trenches gazing motionless towards the Russian line, sometimes alone, sometimes in pairs, one with the periscope, and the other resting to save his strength for the strike.

A boy's eyes gazed hard and cool towards the enemy trench. From the corner of his mouth came the whisper: 'Nothing doing here.' Two men who were walking, bent and stooping, along the trench towards us, straightened a little as they passed and in the same moment a sharp report rent the sodden air like the crack of a whip. We ducked. Who wanted to die in the dreary grey of a wet morning? Again the deathlike silence came down and we went on our way. But the sniper in the front line at Narva stood motionless, gazing into the coming day.

We met more of them farther along the line, this time two boys from the Siebenburgen, Rudolf aged nineteen and Michael aged twenty-four. We talked to them about their homes, about the war, about everything. Rudolf's father was a huntsman, his brother a huntsman, and he had it in his blood. Put a gun in his hand and his eye looked for a target. Michael had seen his first hunt while still a boy. Now they were back at the butts, but here the quarry fired back. 'We must only fire occasionally and we have to hit them when we do,' Rudolf said, his eyes alight, 'otherwise we give ourselves away.'

A Red Army man showed himself in the opposite trench. A quick aim, then 'Crack!' and he fell forward. A hit? – or had he merely ducked for cover? More hours of waiting and then at last a target. Another shot. The man on the other side stopped and tilted over backwards. A pencil mark on the stockade – that one counted. I asked what they thought about as they stood there crossing them off one after another. 'Only that there's one more gone; one less to hold a rifle,' they said.

Sometimes the 'opposite number' across the way spotted one of our snipers. Where was he? Once the target was sighted, a duel developed in which every conceivable trick was used, every scrap of cunning. A whole magazine might be fired off from a feint position, then a quick dash back to the old stand to see if the enemy would reply. Once the position was located it was usually the end of the duel. Sometimes it was our man who lost. Then another took his place as the eyes of the forward line, behind whose vigilance the others could afford to relax.[5]

In the Second World War, as in other wars, the qualities that made a good sniper were fiercely debated, but there was general agreement that the hunting instinct was essential. One sniper authority, Captain Christopher Shore, believed that men who were indifferent shots on the range could make better snipers than Bisley-standard marksmen, as long as the hunting spirit was present:

During the training of snipers there were many instances of men being excellent target shots but failing in the role of sniping shots. Psychologically, the difference between resting on a placid firing point and shooting in war, even if one is in a comparatively safe position oneself, is the greatest possible contrast. Even in hunting the difference is most marked and I have had personal experience of good target shots failing lamentably in the hunting field with its nerve tensions resultant of growing excitement.[6]

In the light of the great strain placed on the sniper, Shore stressed the importance of working in paired teams:

The sniper has a most difficult job, and his battle is a personal, or near personal one; his courage and guts he must have, is of the lonely type; his life is dependent not only on his skill as a shot, but on his qualities of patience and self-discipline, his field craftsmanship, his confidence; and his burning, controlled enthusiasm. And he carries the terrific responsibility of knowing that he can be instrumental in saving the lives of many of his comrades, and he is well able to influence the morale of the whole battalion.

A lone sniper can do much – there are occasions when a sniper working entirely alone can do a more effective job than two snipers working together – but generally speaking the advantages of a sniper team are manifold. One of the salient advantages is elementary psychology – the presence of another man is a terrific booster of morale. Sniping is a nerve-racking game; and the strain tells on even the toughest fibred of men, albeit unconsciously in some cases. Two pairs of eyes are better than one, and many observation reports handed in by sniper teams must have been more comprehensive, reasoned and clear than had a single sniper been out in front.

It was essential in the 'team' idea that the two men forming it were not antagonistic. To send out a sniper pair who were at loggerheads would have been fatal, not only to themselves, but to the job on hand. That fact was clearly understood by everyone.

The team, well equipped, could certainly travel lighter than a lone sniper, and a pair could be certain of their kills since the man who was not taking the shot could have the target covered and under the closest scrutiny either with binoculars or telescope. In certain cases it was possible for the second man, or observer, to put the rifleman on a target he could not really, strictly, define in his telescope sight; for example, a Hun extremely well hidden in the shadow of bushes who defied the power of the TS [telescopic sight] but could be clearly seen through the penetrative telescope could be KO'd by 'team work', the observer indicating the target to the sniper, giving him the fire order and watching the result through the telescope.[7]

Most troops, when captured, expected to be treated according to the rules of the Geneva Convention (although it was widely abused in the Second World War), but snipers could expect to be shot out of hand. The general anti-sniper attitude of most armies was summed up by the distinguished US war correspondent, Ernie Pyle, who wrote of sniping in Normandy and American attitudes towards German snipers:

Sniping, as far as I know, is recognized as a legitimate means of warfare. And yet there is something sneaking about it that outrages the American sense of fairness. I had never sensed this before we landed in France and began pushing the Germans back. We had snipers before – in Bizerte and Cassino and lots of other places, but always on a small scale. There in Normandy the Germans went in for sniping in a wholesale manner. There were snipers everywhere: in trees, in buildings, in piles of wreckage, in the grass. But mainly they were in the hedgerows that form the fences of all the Norman fields and line every roadside and lane.

It was perfect sniping country. A man could hide himself in the thick fence-row shrubbery with several days' rations, and it was like hunting a needle in a haystack to find him. Every mile we advanced there were dozens of snipers left behind us. They picked off our soldiers one by one as they walked down the roads or across the fields. It wasn't safe to move into a new bivouac area until the snipers had been cleared out. The first bivouac I moved into had shots ringing through it for a full day before all the hidden gunmen were rounded up. It gave me the same spooky feeling that I got on moving into a place I suspected of being sowed with mines.

In past campaigns our soldiers would talk about the occasional snipers with contempt and disgust. But in France sniping became more important, and taking precautions against it was something we had to learn and learn fast. One officer friend of mine said, 'Individual soldiers have become sniper-wise before, but now we're sniper-conscious as whole units.'

Snipers killed as many Americans as they could, and when their food and ammunition ran out they surrendered. Our men felt that wasn't quite ethical. The average American soldier had little feeling against the average German soldier who fought an open fight and lost. But his feelings about the sneaking snipers can't be very well put into print. He was learning how to kill snipers before the time came for them to surrender.[8]

Such views were held throughout the American forces, and even General Omar Bradley was known to have privately condoned the killing of captured snipers. Sergeant Harry Furness, an experienced sniper who served in Normandy, was well aware of the dangers: 'All snipers (on both sides), if captured, were shot on the spot without ceremony as snipers were hated by all fighting troops; they could accept the machine-gun fire, mortar and shell splinters flying around and even close-order combat, but hated the thought of a sniper taking deliberate aim to kill by singling them out.'[9]

Not surprisingly, snipers did not subscribe to the common prejudice that they were just assassins in uniform. Captain Shore mounted an eloquent defence of the sniper and the qualities required for success:

The average American or Englishman is averse to killing and many, having been forced to kill, suffer from remorse. But it is a fact that a sniper will kill with less conscience-pricking than a man in close combat. Personal feelings of remorse or questioning of motives will slow down a man's critical killing instinct and the sniper who allows himself to fall into such a train of thought will not last long. It is imperative to look upon the killing of an enemy as swatting a fly, an unthinking automatic action. Two things only should really interest the sniper: getting the job done and getting away unscathed.

To become accustomed to 'sniper killing' is not so difficult or hard as close-quarter killing. A man dies more slowly than the average person thinks; he often grins foolishly when he's hit, the whites of his eyes roll upwards, death sweat gleams on his forehead and he sags to the ground with a retching gurgle in the throat – and it is difficult to hear that gurgle without emotion. The sniper is usually spared all this unless he's quick enough to get his binoculars on his victim, or the 'sniped' is at such close range that the telescope sight will give him all the motion pictures he needs.

Sniping is not the vague, haphazard shooting of the unknown in a sort of detached combat. It is the personal individual killing of a man in cold blood, and is an art which must be studied, practised and perfected. I often heard it said that a sniper should be a man filled with a deadly hatred of the Hun, or enemy. But I found that the men who had a seething hatred in their hearts for all things German, such as those who had lost their wives and children and homes in blitzed cities, were not the type to make good rifle killers. The type I wanted was the man of cold precision, the peace-time hunter who had no hatred for his quarry but just a great interest in the stalk and the kill.

When one is in position for a shot there came an 'inner freezing'; the breathing was not quite normal; the hearing sense was magnified and there came too that sense of excitement which all hunters know and which results in an unconscious nerve-hardening, and once the Hun was in the sight and the pointer steady at the killing spot there was no qualm of conscience about hitting him or taking life. The true hunter is never a butcher; he does not desire to kill for killing's sake, but there is something elemental in the stalk and the slaying which swamps every other feeling and makes the heart and brain exultant, and filled with action-elation.

There are still many unenlightened people who think that sniping is 'dirty', horrible', 'unfair'. I maintain that it is the highest, cleanest game in war. It is personal and individualistic; a game of great skill and courage, of patience and forbearance. And if one loses to a skilful opponent one should never be conscious of it. Death-night to the sniper's quarry should always be of a tropical, sudden intensity.[10]

German snipers were a constant problem, and although their skills were highly regarded by the Allies, they too made mistakes. In one instance, the cardinal sin of smoking while on active duty cost a German sniper his life. Corporal Arthur Hare was a distinguished British sniper whose military career was recorded in novel form by Barry Wynne. Corporal Hare and his observer, Private Packham, had been trying to eliminate a German sniper who had caused their battalion repeated casualties. Wynne recounts how they had little success until the German made his fatal error:

The following day also dawned bright and beautiful, and the shell-hole was once again occupied by the [British] snipers, but all to no avail. It was three days later that their first chance came.

It was mid-afternoon. Perhaps the German had become careless. It was Arthur's turn to be studying the distant wood and the little hill to the west. His body stiffened and Packham instinctively looked up.

'Here, come up and have a look.'

Packham grabbed his rifle and climbed the dry earth wall of the crater. Gradually, like a snake moving in slow motion, he eased himself to the lip of their shell-hole. Without stirring so much as a leaf of their protective screen, he slipped his rifle out through the furrow. Removing the leather cap, he stared through the telescopic lens.

'Where?'

'See that dead tree standing to the right of that hayrick, just on the edge of the wood? Two fingers from there to the right.'

Slowly, Packham swung his rifle. The slightest movement

presented the viewer with a blurred image of rushing ground, like a hastily panned movie camera. He held steady and began to focus with minute attention upon the foliage at the bottom of the trees.

'No, can't see a thing,' he said, two minutes later. 'What am I looking for?'

'Smoke,' was the laconic reply, 'the bastard's having a fag.'

Packham chuckled. Most snipers smoked. It was strictly against orders, but it helped pass the time and steadied the nerves.

'Got it!' Packham rasped, 'you're bloody right. Can't see him though.'

'He'll move, in time,' Arthur answered with firm conviction.

'Whose going to have him?' Packham queried, softly.

'You fire,' Arthur replied, 'I'll watch and make sure you get him.'

The minutes dragged on and became a quarter of an hour. The effort of remaining so perfectly still began to tax their patience. Three days they'd spent waiting for this one, but, well trained, their adversary eluded them.

It was the following morning that the German gave himself away. From the moment that the sun had peeped across the meadow, Packham and Hare had been waiting. The Jerry must have overslept for he was late in getting into position. At 6.10 precisely, his head and shoulders appeared for a second or two framed in an archway in the foliage.

Packham fired a single shot. All that Arthur saw was the German's rifle pirouetting in the sky.

'Got him,' he said with conviction.[11]

A crack shot with many rifle championship awards to his credit, Harry Furness served as a sniper with the Green Howards during the Normandy landings. After recovering from wounds sustained by German mortar bomb splinters, he transferred to the Hallamshire Battalion of the York and Lancaster Regiment, where he served as a sniper for the rest of the war. Here he recalls an incident that required quick thinking:

It was very early a.m. and barely light as two of us from the sniper section came out from an orchard onto a small heavily cambered roadway. We were on our way forward to find a suitable position among the hedgerows from which we could operate that day. We went along very quietly for we knew the area was full of German soldiers, for at that period the battle situation was very fluid and there wasn't an actual front line.

Suddenly, we heard the sound of a fast approaching motorcycle; my mate ran across to the opposite side of the road and lay in the ditch, as I too got down. As soon as the German despatch rider appeared in sight, I fired. He spun off his bike and fell on his back onto the roadway; his motorcycle continued on into the ditch. I rushed towards the German soldier who was wearing a helmet with goggles and a long waterproofed coat. It was very obvious to both of us that we would have to move very fast to retrieve anything useful for our battalion intelligence officer, for the sound of our shot would alert any German soldiers in the area. I just had time to take the map case which had been slung around his shoulders, then we quickly moved away. As it happened, each of us thought we had been the only one to shoot, as it had sounded only as one shot. But, in fact, we had both fired precisely at once, as the two entry wounds in his chest could be clearly seen.[12]

On a subsequent mission, Sergeant Furness had the opportunity to eliminate a high-value target:

I can recall one occasion when I had good reason to think that my quick snap shot had hit a senior officer who was being shown our British front-line area. He was with a group of others, and actually didn't present much of a target as they were behind some cover when I first spotted them some several hundreds of yards away. The officer who caught my attention was in the centre of the group using field-glasses, when I chanced a very quick shot at him. Following my shot there was immediate commotion as they dropped behind cover, so I could not observe more.

I realized they would already be searching for my position with their binoculars, so I started to withdraw slowly backwards. But a furious retaliation barrage came down around my location very quickly indeed. It seemed to me they were firing everything they had, with machine-guns, mortars and 88mm shell fire. The barrage they put down was continuous, with explosions and the screech of flying metal all around me. Although I lay flattened out on the ground, several times I was lifted by the concussions. I had my mouth wide open with my arms wrapped around my face, for we had been told that more damage by explosions could occur if we had our mouth closed tight. I had been through many barrages previously but nothing on the scale of this never-ending retaliation, so I knew I had good reason to believe I had shot down a very senior officer.[13]

Peter Young was the CO of No. 3 Commando, which had landed on D-Day and been responsible for defending the open left flank of the British invasion force. In the hedgerows of Normandy, Young upheld the traditional Commando interest in sniping:

It was on the 15 June that we inaugurated the sniping season. We had selected and trained snipers when we were in England. The men chosen were the best shots, but this, oddly enough, did not work at all, for we soon found that many of the best marksmen had not the temperament for the lonely work of a sniper. Many of the men who enjoyed sniping were by no means remarkable shots, but they would creep up so close on the enemy that they could not miss! In fact in this game stalking is as important as shooting.

An enterprising character in 1 Troop, Trooper Fahy, made himself a camouflage suit from denim overalls, hessian strips and odd pieces of material that came to hand. Then nothing would do but he must go off and try his luck. He and Trooper Needham, a cool, resolute Irish Guardsman, similarly attired, showed me their suits and persuaded me to let them loose. I sent them down to the Longuemare crossroads and they were back within half an hour, having shot two Germans. The enemy in the Bas de Breville opened up with a machine-gun, but there was plenty of cover by Dead Horse Corner.

The snipers had a good day on 25 June when TSM Edwards of 3 Troop and Lance-Corporal Osborne MM went out together from La Grande Ferme about 12.30. For about two hours they crept about in hedges and orchards, but they were quite unable to find any Germans and returned to the farm, where they were met by a sergeant of 4 Commando and decided to try again. Leaving by the east gate, they made their way along the hedges to a bomb crater. While the other two watched, Osborne moved out into the cornfield and proceeded to shout and wave his arms as if he were leading a section of men forward. After two minutes he reported enemy movement and the other two spotted a group of Germans watching Osborne's antics from a gap in the hedge. They fired together. Two men fell. An enemy machine-gun opened up, but the snipers returned to our lines unscathed.

The sniping went very well; indeed, our bag gradually increased to quite respectable proportions, and that without loss to ourselves. In 3 Troop Corporal Hanson and Trooper Hawksworth were the star performers; with TSM Edwards and Corporal Osborne MM, they once accounted for five Germans in one day. After this the enemy became so careful that it was hard to find targets.[14]

Snipers were expected to use their initiative, and their skills in acquiring intelligence of enemy dispositions was often as important as their shooting ability. Harry Furness relates how two snipers from the Hallamshire Battalion managed to turn a dangerous situation to their advantage:

> Two of our snipers were working well forward of our front line, and after shooting at a group of enemy soldiers began moving to another position in case they had been observed from muzzle blast, etc. Suddenly, the entire area became alive with German infantry, so they quickly moved into a shell-damaged house, to try and hide until the soldiers passed them by. They went upstairs and just had time to get into the rafters when a section of German infantry also moved into the house. After only a hurried glance in all rooms, the Germans settled down in the lower ground floor, making it impossible for our two snipers to escape. They had no option but to settle themselves on to the exposed rafters and wait, but in the meantime they used their binoculars and scout telescope to observe the German units in the area.
>
> It was a couple of days before the German infantry moved on. The snipers had taken it in turns to snatch short rests in between keeping out of sight, but they did manage to compile a very useful log on the identity and equipment carried by the German units, which they were able to bring back to our battalion HQ. Meanwhile it had been thought by our battalion that they were missing in action, probably dead, as snipers were always promptly dealt with on the spot. But it clearly demonstrated, once again, that it takes a lot to keep a good Hallamshire down.[15]

After the Allied breakout from the Normandy beachhead – culminating in the near destruction of the German Army at Falaise – there were few opportunities for sniping, so fast-moving were military operations. But towards the end of September 1944, German resistance increased and open fighting ended as the two sides faced each other in a line stretching through Holland and Belgium and along the Franco-German border. It was at this time that sniping reasserted itself.

In Holland, German and British positions could be separated only by a canal or other waterway. Christopher Shore relates an incident of sniping at very close range:

> The battalion was holding one bank of the Nederweet Canal in Holland with the Germans on the opposite bank. The distance between the combatants was only about twenty-five yards and it was possible for the snipers to hear the Germans speaking quite distinctly. There was a very high bank on each side of the canal and although the battalion snipers waited patiently for hours on end Jerry was very careful, and remained in the safety of his own towering bank.
>
> But one afternoon the snipers' patience was rewarded, since for some reason or other the Germans decided to have a celebration, and proceeded to get really drunk. The first Hun to be accounted for had a bottle of wine to his lips and was in the act of taking a long draught. Perhaps that is as good a way to die as any! The shot caused a little consternation in Jerry's camp, but not much. The interpreter from the battalion intelligence section was alongside the sniper who had 'bumped-off' the imbibing one, and he was delighted to translate the resultant conversation for the edification of the snipers. Immediately after the shot had been fired and the German with the bottle killed, a wine-thickened voice bellowed, 'Who in the name of Venus fired that shot?' A reply in a similar voice was, 'I don't know, but who's the silly ───── who's been shot anyway?'
>
> The afternoon party was very costly for the Hun, for before nightfall the snipers had killed five. It was really too easy. One wonders what the German platoon commander thought about it all next morning when he awakened with probably a damned bad head and found that his platoon had indeed been sadly depleted in strength.[16]

In a further sniping incident in Holland, Harry Furness demonstrated that patience and close observation bring their rewards:

One of our rifle companies holding a position near a waterway in Holland had suffered casualties from accurate sniper fire. I was sent from our section to see if I could pinpoint the sniper's position. After asking a few questions from the company officers, I searched around for a suitable vantage point to observe a wide area, and chose to go into the damaged roof rafters of a high house flanking the rifle company. But over a long time no further shots were fired, and I thought it possible he had already moved to another part of the front-line area. However, I continued to search for any signs of him using my 20-power scout telescope.

All the houses seemed to have suffered some damage, either from shelling or mortar bombs, and one house had broken windows and damaged exterior wooden shutters. It looked to be deserted, but now and again one of the shutters moved a little as the wind caught it and banged it against the wall. It was too far away for me actually to hear it, but I could clearly see it sway as the wind caught it. But for someone inside the house it would have sounded louder and might attract someone's attention.

From my position the interior behind all the windows was dark and at first nothing could be seen. I must have been watching the area and that swaying shutter for hours when suddenly I caught a quick, little movement, which draws the eye. I switched from my scout telescope to pick up my rifle and I watched intently through my telescopic sight, and I saw what seemed to be a hand reaching to catch hold of the edge of the loose shutter. I fired immediately into and near the edge of the shutter, and even through the rifle recoil I felt sure I had seen an arm slide down and bang on the window sill and then disappear.[17]

Subsequently, a British patrol searched through the building and returned with a semi-automatic G43 rifle with telescopic sight, which they had found on the floor alongside a dead German. Furness had shot the German sniper.

In certain areas of the line, 'freelance' rifle fire was discouraged for tactical reasons. Even so, it was hard to stop determined individuals from taking a shot at the enemy when a good target presented itself. Captain Shore's battalion was holding a position near the German border, and he describes an instance where a sergeant in his battalion had recently acquired a German sniper rifle and was itching to use it on its former owners:

> Our policy here was one of strict 'non-firing' in order to keep the Hun guessing and make him nervy, which certainly proved to be the case in the end, but the sight of a German gunner OP officer, easily recognizable at the range [250–300 yards], calmly surveying our village through his binoculars proved too much for the sergeant one morning. He was observing from the loft of a house in his platoon area and saw this Hun officer sitting there in full view and smirk-suggestive in his arrogance; the sergeant became so incensed that ignoring all rules and regulations he took his German sniper rifle, which so far as anyone knew he had never fired or zeroed before, and very calmly shot the officer slap through the head! There was no counter-battery fire so the assumption must be that the desired result had been obtained![18]

Charles Askins, the indomitable American shooter, was another man who refused to let officialdom come between him and sniping the enemy. In a long career, Askins had been a Texas ranger, big-game hunter, small-arms author and soldier. Too old to be able to serve in an infantry unit, Askins received a commission as an officer in a combat salvage unit. Assigned to north-west Europe, he covertly managed to get to the front line for some direct shooting experience.

Following the US victory at the Battle of the Bulge, American forces began the slow advance into Germany. Early in 1945 the Americans and Germans faced each other across the River Rur, which divided the town of Düren. Askins sensed a possibility for some hunting. Armed with an M1 rifle, and taking along with him a driver of Greek origin named Papalexiou, Askins took up a position in a ruined house overlooking the German lines on the far side of the river:

There were likely targets during the morning but the sun was all wrong. It was shining in my eyes and although I had blackened my sights I did not want to take a chance of the rays in my eyes maybe resulting in a miss.

Midday came, and went, and finally the sun was behind us. In the room was a very comfortable divan. The Big Greek and I hauled it around and moved it against the window. The latter had long since been shattered and offered no hazard. I got into a comfortable position on the couch, the rifle rested over the padded back of the long seat.

Very directly a Wehrmachter sauntered down to a fringe of bushes within thirty feet of the river, hung his rifle on a nearby branch, slipped out of his greatcoat, hung it on another bush, and pulled down his pants. That didn't surprise me in the least. I had been watching the performance of this particular Heinie for three days and I could set my watch by his bowels. This was his regular latrine and I think he was of the opinion that he could defy the 'Amies' by taking his crap in plain view of a couple of hundred.

He was behind the screen of bushes but I could see the outline of his hunkered body very plainly. I held for his big middle and meticulously squeezed off the shot. On the recoil of the M1, which was piddling light, I promptly snapped in another round. The Big Greek, who was watching through the binoculars, turned to me grinning, 'Both shots hit him, I saw the body jerk.'[19]

Askins and Papalexiou hurriedly vacated their position, less to avoid any German retaliation, but rather to escape recriminations from US infantry authorities, incensed that their section of the line had been taken over by these freelance snipers. But the success of the first mission brought them back to Düren three days later, in search of more targets. Adopting a new position, the two Americans looked across at buildings which had survived the worst of the Allied air bombardments and might be housing Germans. Of these buildings, Askins wrote:

One appeared to be a school, or possibly it was a monastery, for there were a whole series of windows which faced towards the river and all of which, of course, were shattered. In watching this structure I was struck by the fact that there was something new at the window in the very top corner of the third floor. I studied this window and whatever it was in the deep shadow behind the opening.

It finally occurred to me that this was a big spotting scope. It had the right size for about a 40×, set on a heavy tripod, and as I kept my eyes glued to the scope, I was delighted to see that sitting comfortably behind the eyepiece was a Kraut soldier. It was obvious that this was either an artillery observer or a mortar observer. As the sun marched in its usual fashion, the light into the room grew stronger. At the elbow of the observer was a field telephone and there was a field table, and watching keenly, I saw that the man did not give too much attention to his glass. It was obvious he had been there for days, maybe even weeks, and was pretty bored with the job.

Here was a made-to-order target but the problem was that he was a full 100 metres beyond the street intersection which lengthened my yardage to about 380 metres, a hell of long way for the old M1 and its puny '06 round. Still I was delighted with the prospects.

At 1600 hours, I could see the man with enough

definition to make ready the shot. He was slouched down at
one side of the big glass and I was concerned that I might
hit the telescope and not the human. But about that time
the SS trooper stood up. He was a big man, at least six feet
in height with broad shoulders. I could see him above the
window frame from his thighs upward. I held at the top of
his head and touched off the shot. I looked at my driver. 'He
disappeared when you shot,' said the Big Greek. 'Watch the
room', I enjoined, 'and let's see what happens.'

Soldiers bustled into the room. There were four of them.
I fired just as fast as I could find a target in my sights. I let
go five rounds just as quickly as the trigger could be pulled.
'They're going down,' the Greek was excited now.

'Come on,' I whispered, 'let's get the hell out of here.'
We scrambled down our back stairs and made a dash for the
alley. It was not uncommon for an occasional shot to mar the
peace of a quiet sector on the front but six shots were too
many. It would alert all our people on our side of the river.
Battalion HQs would be on the phone wanting to know
what the hell they were shooting at. We piled into our jeep
and ran like hell for camp.[20]

Apart from the acute observational skills and marksmanship
displayed by Askins, the account is noteworthy for his disparaging
comment on the carrying power of the .30-06 round used in the
M1 rifle. Most other snipers were content with this round at the
ranges cited by Askins (no more than 400 yards), but this
criticism may have reflected a big-game hunter's predilection for
powerful magnum-style rounds.

In the British Army, long-range shooting seemed not to play
a large part of the sniper's training and outlook. The British No.
4 (T) sniper rifle was capable of accurate fire at ranges of up to
1000 yards, but snipers preferred to work at much closer ranges.
In the following extract written by Christopher Shore, a distance
of 800 yards was considered beyond the reach of an experienced

sniper section. Shore describes the activities of the snipers, operating close to his own company, and their simple solution to the problem:

Day after day they saw a German observer who was using a high tree as an OP [Observation Post]. This tree was some eight hundred yards away from the sniper post and the intervening country was barren and so devoid of cover that it was decided it was impossible to get anywhere near enough to make sure of a kill. This Hun was a very cocky individual, swinging up into a tree, having a good look round with his binoculars, and then swinging to earth again for a rest and a smoke. Naturally it was not long before he became known to the snipers as 'Tarzan' and his monkey-tricks were looked upon with a mixture of amusement and chagrin.

Having decided that they really couldn't do very much about it the section leader asked the Forward Observation Officer of artillery who was with their battalion headquarters if he could help to liquidate 'Tarzan'. The officer said that he would be delighted to try, and a field telephone was laid from the snipers' post back to the battery.

The artillery officer went up to the post next morning, had a good look at 'Tarzan' doing his acrobatics in the tree and then became very engrossed in his map, plotting the target. Finishing that task he rang up the battery, gave them 'the dope' and ordered one round to be fired. And that one round scored a direct hit on the tree! When the snipers looked again there was no sign of 'Tarzan'. After witnessing such shooting the snipers were forced to admit, laconically, 'Not bad for the first attempt'; but really they were very impressed, and from that time onwards the artillery officer was treated with great respect. After all to snipe with a 25-pounder is great work! Later, when the battalion moved forward, 'Tarzan's' helmet, bearing a great shrapnel gash, was picked up. And in it was part of his head![21]

Artillery had proved highly effective in dealing with enemy snipers and observers on the Western Front during the First World War, and continued to do so in subsequent conflicts. Indeed, an old US Marine Corps saying states that the best sniper weapon in the world is a 155mm howitzer.

In the Pacific, the Japanese were masters of camouflage and concealment, and although far from brilliant marksmen, their snipers caused repeated problems for the Allies. Limitations of terrain, particularly jungle, ensured that ranges were short, but as in the trenches of the First World War, pinpoint accuracy could be vital. Until the Allies instigated serious counter-sniper policies, the Japanese sniper remained a potent threat. The journalist Richard Tregaskis was stationed on Guadalcanal in the Solomon Islands, where the US Marines were desperately fighting the Japanese for control of the island. He reported his first contact with a Japanese sniper while on an expedition to the nearby island of Matanikau:

> More Jap .25s opened up ahead; a storm of firing broke and filled the jungle. I dived for the nearest tree, which unfortunately stood somewhat alone and was not surrounded by deep foliage. While the firing continued and I could hear the occasional impact of a bullet hitting a nearby tree or snapping off a twig, I debated whether it would be wiser to stay in my exposed spot or to run for a better 'ole and risk being hit by a sniper en route.
>
> I was still debating the question when I heard a bullet whirr very close to my left shoulder, heard it thud into the ground and then heard the crack of the rifle which had fired it. That was bad. Two Marines on the ground ten or fifteen feet ahead of me turned and looked to see if I had been hit. They had evidently heard the bullet passing. That made up my mind. I jumped up and made for a big bush. I found it well populated with ants which crawled up my trouser legs, but such annoyances were secondary now.

The origins of British sniping in the First World War: an officer of the Argyll and Sutherland Highlanders casually snipes across no man's land in the Bois Grenier sector, March – June 1915. (Imperial War Museum)

Recruits undergo training at a British sniper school, July 1916. The introduction of sniping schools at army level in 1916 made sniping more professional. (Imperial War Museum)

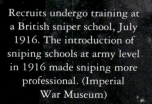

Main picture: A British sniper looks through his telescopic sight from a house in Caen, July 1944. The sniper would normally fire from deep within the house, and not at the window as shown here in the photograph. (Imperial War Museum)

Right: A French soldier prepares to fire at the enemy using a sniperscope attachment, August 1916. Although it provided the firer with protection, accuracy was poor. (Imperial War Museum)

Above: A sergeant of the Lovat Scouts trains snipers for deployment in North-west Europe. Recruited from Highland gillies, the Lovat Scouts were famed for their observational and stalking skills. (Imperial War Museum)

Right: Sergeant Harry M. Furness, who served as a sniper with the Hallamshire Battalion of the York and Lancaster Regiment in North-west Europe, 1944–5. (AFPU)

Left: A US soldier in Korea scans Chinese positions with a spotting scope. Sniping became increasingly important in Korea as the war bogged down into attritional stalemate. (National Archive)

Main picture: A Soviet sniper is depicted taking up a firing position in the snow of the Caucasus mountains. Large numbers of sharpshooters and snipers were active in the Red Army. (National Archive)

Left: Armed with an M40 rifle, a US Marine sniper stands in a paddy field in Vietnam. Based on a civilian hunting rifle, the bolt-action M40 proved popular with US snipers. (Kate Chandler)

Right: A sniper in the 7th US Marines, Thomas D. Ferran was credited with 41 kills during his 12-month combat tour of Vietnam. (Thomas D. Ferran)

Left: A British Royal Marine sniper zeros his L42A1 rifle while on active service in Northern Ireland. The 20-power spotting scope is on his right. (Private Collection)

Below: A British sniper armed with L96 sniper rifle (second left) returns from a four-man patrol in Northern Ireland's border country. (Private Collection)

Top : Two British soldiers use their L96 sniper rifles to good effect, displaying their trophy while on a break from a training exercise in Africa. (Private Collection)

Right: Anglo-American co-operation: A US Marine sniper armed with an M40A1 rifle (right) is supported by a British Army sniper armed with an M16 (left). (Private Collection)

Below: A British sniper fires a .5in Barratt heavy sniper rifle; it can penetrate light armour and is capable of hitting targets in excess of a mile. (Private collection)

Above: Correct sniper deployment: wearing a gillie suit and making good use of natural camouflage, the sniper is positioned at the base of the main tree. (Private Collection)

Below: Dressed in trademark gillie suits and armed with L96 sniper rifles, students on a sniper instructor's course pose for the camera. (Private Collection)

The sniper who had fired at me was still on my track. He had evidently spotted my field-glasses and taken me for a regular officer.

I searched the nearby trees, but could see nothing moving, no smoke, no signs of any sniper. Then a .25 cracked again and I heard the bullet pass — fortunately not as close as before. I jumped for better cover, behind two close trees which were surrounded by ferns, small pineapple plants and saplings. Here I began to wish I had a rifle. I should like to find that sniper, I thought. I had made an ignominious retreat. My dignity had been offended. The Matanikau sortie had become a personal matter.[22]

Later in the Guadalcanal campaign, Tregaskis was reporting on the activities of the celebrated 'Edson's Raiders' when he came under further sniper fire:

I was sitting on the side of the ridge that looks over the valley where our tents are located. A throng of Zeros were dog-fighting with our Grummans in the clouds and I was trying to spot the planes.

Suddenly I saw the foliage move in a tree across the valley. I looked again and was astonished to see the figure of a man in the crotch of the tree. He seemed to be moving his arms and upper body. I was so amazed at seeing him so clearly that I might have sat there and reflected on the matter if my reflexes had not been functioning — which they fortunately were. I flopped flat on the ground just as I heard the sniper's gun go off and the bullet whirred over my head. I knew then that his movement had been the raising of his gun.[23]

Fortunately for Tregaskis, a more competent sniper would not have made the mistake of moving so obviously and would probably have hit his target. Angered that he had again been the

subject of a sniper's personal attention, Tregaskis was gratified
when several Marines fired into the tree where the sniper had been
hiding.

Tree positions were a feature of Japanese sniping; the obvious
dangers of easy detection and elimination were considered worthwhile if a single GI could be killed. A US intelligence bulletin of
November 1942 alerted its readers to the methods used by the
enemy:

> Cleverly hidden Japanese snipers proved very troublesome to
> the US Marines. A Marine sergeant reported our biggest
> problem was in locating and destroying snipers.
>
> They were concealed in trees, bushes and buildings.
> Time and again, our forces passed through an area and we
> were shot at from the rear. A second Marine officer said that
> the Japanese used a large number of snipers, well camou-
> flaged. 'They shot at us from the top of coconut trees, slit
> trenches, garden hedgerows, from under buildings, from
> under fallen palm leaves,' he explained. 'One sniper, shot
> down from a tree, had coconuts strung around his neck to
> help conceal him. Another in a palm tree had protected
> himself with armour plate. Our Thompson [sub-machine-
> guns] and BARs [Browning Automatic Rifles] proved to
> be excellent weapons for dealing with snipers hidden in
> trees.'
>
> The snipers sought especially to pick off officers and non-
> commissioned officers who wore insignia and markings
> indicating their rank. The Japanese placed snipers on the
> flanks of their positions and weapon emplacements.
>
> A large number of Japanese wore green uniforms and
> painted their faces and hands green so they would be hard to
> see among the green vegetation of the islands. They also
> wore camouflage nets and wood fibre strands garnished with
> vegetation. Japs wearing these were hard to see, even at fifty
> yards, if they were still.[24]

The particular advantages enjoyed by the Japanese sniper over his American counterpart were explained in an article in the *US Marine Corps Gazette* of January 1945:

There are certain advantages in jungle warfare which are invariably entered on the black side of the defender's combat ledger. The defender is usually able to select the position he will defend, dig in, get up his supplies, site and camouflage his weapons, conceal his personnel, establish his security, and wait for the attacker, all according to a previously prepared plan. The attacker is burdened with the problem of locating the defender, feeling out the position, and blasting his way through it. In the process of doing so, he cannot expect to employ cover and concealment as effectively as the defender because he must move. In moving, he must expose his personnel. Thus the well-concealed enemy sniper is automatically provided with an abundance of targets without exposing himself.

The Japs are now showing a tendency toward .30 calibre small arms in an effort to secure penetration and 'brush cutting' qualities. However, the majority of Jap soldiers and practically all snipers still use .25 calibre weapons. Snipers are usually equipped with rifles, carbines, or Nambu light machine-guns.

The .25 calibre weapon will not cut brush or penetrate like a .30 calibre weapon. It will, however, penetrate our helmet at ranges of 150 yards or more. This is penetration enough in most cases. Its most annoying characteristic is that for all practical purposes the Japanese small-arms powder is actually smokeless and generates little muzzle blast when used in a .25 weapon. Thus a sniper can 'hole up' or 'tree up' with any of his three small arms, and although he may fire considerably, we will seldom, if ever, locate him by smoke or muzzle blast.

At the beginning of hostilities (and in many cases even now) the Jap was generally a better woodsman than our

Marines. He was meticulously trained and equipped for
jungle warfare long before Pearl Harbor and he had received
the training and used the equipment in the jungle. The
training had been realistic and tough. Jungle techniques had
been developed and proven while we were parading in blues
at the Marine Corps Base. The Jap sniper had learned to live
as successfully as an animal in the jungle while our boys were
enjoying a standard of living they had been raised to regard
as a right, rather than a privilege.[25]

The importance of employing first-rate marksmen to deal with
difficult targets was underscored in a further account from Richard
Tregaskis. One of the hardest targets to hit is a fast-moving man,
especially when he is running for his life. A whole platoon of
infantrymen will often fail to hit the escaping man. This narrative
shows how eventually it took a crack shot to bring down a
Japanese soldier who had slipped past a line of three American
flame-thrower tanks during fighting on the Tenaru Front:

> The Jap continued to run. He was heading for the beach. All
> along our front line, rifles banged and machine-guns clat-
> tered; the tracers arched around the running Jap. Then the
> Jap sank into the underbrush, took cover, and Colonel
> Pollock shouted: 'Don't shoot. You might hit our own
> tanks.'
>   The Jap jumped up and ran another forty or fifty feet
> towards the shore, then sank into cover again. Despite the
> warning, several shots were fired at him. As usual, each
> Marine was eager to kill his Jap.
>   'One man fire,' shouted Captain Sherman. He designated
> a grizzled, leather-faced Marine to do the shooting. I noticed
> that the man wore the chamois elbow pad and fingerless
> shooting glove of a rifle marksman. The Marines told me he
> was Gunnery Sergeant Charles E. Angus (of Nashville,
> Tenn.), a distinguished marksman who had won many a
> match in the States.

We watched Sergeant Angus, as if he were the spotlight star of a play, when the Jap jumped up again and began to run. Angus was nervous. He fired several shots, working his bolt fast, and missed. He inserted another clip of cartridges, fired one of them. But then the Jap had sunk down into cover again.

It was a little disappointing – but only for the moment. The Jap had flopped on the beach. He was evidently heading for the sanctuary of the water, hoping to swim for it. But now he started to get up again – and that was as far as he got. He had only reached a crouch when Sergeant Angus, now quite calm, took careful aim and let one shot go. The Jap sank as if the ground had been jerked out from under him. It was a neat shot – at about two hundred yards.[26]

The key to dealing with Japanese snipers – despite their talented use of camouflage and their stoical endurance – was determination and the adoption of intelligent counter-sniper methods. The Allies were generally slow to develop such tactics, although sometimes the determination alone was sufficient. Russell Braddon, an artilleryman in the Australian Army, witnessed this attempt to eliminate a troublesome enemy sniper:

I was amazed to see a fellow gunner raise a heavy Boys anti-tank rifle to his shoulder, aim high and fire. He was at once flung backwards, whilst the half-inch shell most certainly passed harmlessly into the stratosphere. When I reached him he was rubbing his shattered right shoulder and swearing softly but with that consummate fluency which is the prerogative of the Australian farmer who is perpetually harassed by the cussedness of things inanimate.

'What the hell are you trying to do, Harry?' I asked.

'Get that bloody sniper up the top of that bloody tree,' he replied tersely. It appeared that, fired off the ground, the Boys rifle had not sufficient elevation to hit a tree high up. However, since the sniper fired from behind the top of the tree trunk he could only be shot through it – a Boys rifle was, therefore, essential for the job. We decided to do it together. With the barrel resting on my shoulder, the butt against his own. Harry took a long aim, apparently quite undeterred by the bursts of bullets from all sides, which our stance attracted. I was not in the least undeterred. In fact, as we stood there, our feet spread wide apart to take some of the shock, I was very deterred indeed. Then Harry fired and I was crushed to the ground and Harry was flung against a tree and the sniper toppled gracelessly out from behind his tree, thudding on to the earth below, and our job was done. I left Harry, still swearing volubly and rubbing his shoulder, and crept back to the line of men I now knew so well.[27]

In New Guinea a US Army unit – a battalion of the 163rd Infantry, 41st Division – developed an anti-sniper policy, combining improvisation with the most methodical application. The 163rd had taken over the Musket perimeter in January 1943 and were immediately subjected to a galling barrage of sniper fire from Japanese soldiers hiding in the enveloping jungle. This account, drawn from regimental records and collated by the divisional historian Dr Hargis Westerfield, describes how the Americans overcame the sniper menace:

Naturally, Jap sniper fire plunged down regularly at chow time. Once, we had the stoves set up and we had to leave the holes for food. Sniper fire crackled in on us at any unexpected time in twenty-four hours. And at dusk, ground terrorists would probe into spots that snipers' binoculars had observed to be less protected. Lone men or small patrols would work around our flanks or rear, empty a clip or two rapid fire, then escape.

Beneath the eyes of these killers, life quickly became not worth living unless we could shoot them out of the trees. Grimly the Musket garrison – 1/Bn, Hq. Co. with A,B, and D Co.s and ATs [Anti-Tank] 37mm cannon – developed a system of counter-sniping. Our basic tactics consisted of three main steps — and a fourth which AT 163 added with at least three 37s.

First, we began to deal with the most immediate threat from Jap Perimeters Q–R which lurked in holes 20–30 yards before us. We set up two-man counter-sniper teams in slit trenches on the forward edge of Musket perimeter. While one man quietly scanned the opaque jungle with field-glasses — or the naked eye if he had no glasses — the other man cuddled his well-cleaned rifle and waited. When the Jap shots rang out, the observer carefully spotted the green area where the shots came from. He pointed out the direction of the fire, let the rifleman observe through his glasses. Then the rifleman fired — until the Jap was silent — or Jap fire retaliated close enough to make him lie prone. Thus we secured our forward area.

Second, we sent counter-sniping teams into the trees on the flank and rear of Musket perimeter. To lessen the drudgery and danger of climbing among dead branches in jungle sweat, we set up homemade ladders. Usually we made them of telephone wire with stout wooden rungs.

Once the two-man tree teams were aloft, we got to work. We shot at all trees which seemed to harbour Nippo rifles. When Japs fired, we followed our standing order. All teams returned fire. If unsure of the actual target, we engaged probable Jap trees in the general direction of the popping fire. With our M-1s and 1903 rifles, we shot at 200–400 yards. Many men preferred the '03 rifle; they believed it more accurate than the new-fangled Garand M-1.

Third, we needed still another measure, because manning forward slit trenches with two-man counter-sniping teams was not enough. As soon as we posted sniper teams in trees, we could take the offensive. We could use these teams to guide attack patrols on the ground. We sent out small foot patrols of two–three men. Under direction from tree observers, our patrols shot down snipers or slashed other targets on flanks of Jap Perimeters Q and R.

And the ground patrols also set booby traps – grenades tied to two separate trees and connected by a trip cord attached to the loosened firing pins. These booby traps caused Jap casualties, and once definitely effected our capture of a Jap Bren gun. Evidently the Jap had dropped the Bren gun when they fled from a grenade blast.

When we counter-sniped in these three steps, we carefully secured ourselves from accidentally shooting our 163 men. We briefed all our men on our methods. We located our own sniper trees so that nobody thought we were firing on him. Most important of all, we made it clear that nobody could fire on Jap snipers – except regularly designated counter-snipers.

Fourth, with the arrival of AT 163's 37mm cannon (at least three) we took another step against these hidden Jap killers. When F Co. passed through Musket on 9 Jan., Jess Fallstick noted that the AT 163 gun crews were hard at work, loading 'grapeshot' into the 37mm guns.

Methodically, AT's carefully aimed 37s were topping the jungle trees around Musket perimeter. For without tree cover, no snipers could operate. In BAR-man Fallstick's opinion, the number of trees made the task hopeless, but he admitted that he saw a tremendous number of mangled trees on the horizon. Thus did 163 Inf's 1/Bn counter-sniper the Nip snipers who took sight pictures on us from above Musket perimeter.[28]

The counter-sniper plan pursued by the men of the 163rd reflected their belief in the importance of firepower to suppress enemy sniper fire, and was somewhat at variance with the classic one-on-one approach. None the less, the 163rd system seemed to work and Japanese sniper fire lessened dramatically over the ensuing months. Yet, if nothing else, the sheer size of the American operation was a backhanded compliment to the Japanese sniper.

During the fighting to capture Saipan in June 1944, one

Marine veteran recalls an instance of the classic sniper approach to suppressing enemy fire:

> We were pinned down on the beach at Saipan by a machine-gun bunker. The pill-box commanded a sweeping view of the area, and there was just no way we could get at it. Plenty of our boys had died trying.
>
> Finally one of our ninety-day wonders [an inexperienced officer] got on the horn and requested a sniper. A few minutes later, I saw two old gunnery sergeants sashaying towards us, wearing shooting jackets and campaign hats! As soon as I saw these Smokey Bears bobbing over to us, I figured this could be some show. And it was.
>
> These two old sergeants skinnied up to the lieutenant and just asked him to point out the bunker. Then they unfolded two shooting mats, took off their Smokey Bears and settled down to business. One manned a spotter's scope while the other fired a 1903 Springfield with a telescopic sight rig.
>
> That bunker must have been 1100 or 1200 yards away, but in just a few minutes, with three or four spotting rounds, this old gunny on the Springfield slipped a round right into the bunker's firing slit. One dead machine-gunner. But their commander just stuck another man on that gun. Our sniper shot him, too. After the fourth man bit a slug, I think they got the idea. We moved up on their flank and destroyed the bunker while our snipers kept the machine-gun silent. Then the two gunnys dusted themselves off, rolled up their mats and settled their Smokey Bears back on their heads. And just moseyed away.[29]

The sniping experience of the US armed forces increased as the war progressed. Different theatres of war required different tactics. In the open terrain on Okinawa, for example, the US Army and Marines fought a well-entrenched enemy, and accordingly used armour-piercing bullets to smash through Japanese emplacements and 'rattle the Nips out of their bunkers'. In other areas, where jungle predominated, three-man sniper teams were often employed: one sniper, one observer and a man armed with a sub-machine-gun to provide close-in security. When tree-bound Japanese were encountered at short range, a Thompson sub-machine-gun was often preferred to a sniper rifle; as one Marine observed, a Thompson 'could really mess up a coconut tree'.

The success of sub-machine-guns in jungle counter-sniping is described in this report on the activities of an unnamed US Army sergeant operating in New Guinea:

Almost as soon as this patrol was established, snipers fired on the men. The ingenious sergeant devised successful tactics which depended on the density on the surrounding jungle. Prowling into thick undergrowth, he carried a Tommy gun with 100 rounds. He also carried a canteen on his ammo belt, a hunting knife, and a head net against mosquitoes. Wearing dyed green fatigues, he stained his face and hands with black mud.

When a sniper fired on them, he took an azimuth towards the crack of the rifle, and moved under cover in that direction. When he had closed in on the sound, he lay still until he heard another shot and spotted the exact tree the sniper had fired from. In almost all cases, the sergeant waited until he saw the target Jap with the peaked cap and the long rifle. Then his heavy .45 bullets tore the Jap from the tree. A few times when the Jap was invisible, he slashed bursts into the foliage to kill him.

When scouting for snipers in more open areas, the sergeant carried his M1 for greater accuracy at long range. Although sometimes he took two–three other riflemen with him, he preferred to work alone. He moved from cover to cover towards the firing until he located the Jap he longed for. The sergeant claimed a kill of five Jap snipers and perhaps deterred a larger number from harassing his squad.[30]

The US forces in the Pacific demonstrated an admirable flexibility in the deployment of weapons for counter-sniper measures, employing the maxim, 'if it works, use it'. And even though Japanese snipers continued to claim American lives until the final surrender, their threat was greatly diminished.

The atom bomb brought the Second World War to a close. The initial obsession of the military with nuclear weapons led to a downgrading of the importance of ground forces, but the communist invasion of South Korea made strategists realize that conventional wars would continue to be fought. Similarly, the sniper would also continue to play an important role on the battlefield.

Chapter Six

Containing Communism

The end of the Second World War marked the beginning of the Cold War: more than four decades of hostility between Communism and the capitalist West. Although the two superpowers stopped short of direct conflict, they fought proxy wars, in which the forces of one side engaged an ally of the other. The first significant conflict was the Korean War; the United Nations, led by the United States, fought first the North Koreans and then the Chinese.

Initially, United Nations troops suffered heavily from Chinese sniper fire, and casualties mounted. Both the US Army and Marine Corps were forced swiftly to rethink their casual outlook towards sniping. Lieutenant-Colonel Glen E. Martin described the methods adopted by one US Marine battalion to improvise a sniper policy:

In April 1951, the 2nd Bn, 5th Marines was notified that enough sniper rifles would be on hand shortly to bring the battalion total up to eighteen. We all agreed that the weapons could be used effectively if we could get men qualified to operate them, and then get these men to the right place at the right time.

Frequently, targets at medium ranges and greater could not be handled effectively. They were often beyond effective small arms range, and [yet] too limited in size for economical expenditure of larger calibre ammunition. Moans, groans and profanity proved ineffective against these provocative targets so we turned to the sniper group. Thus it was decided that the battalion executive officer should be given the job of forming a sniper platoon. In selecting personnel for this group it was decided that they should be men who had qualified as experts on the rifle range, could adapt themselves to the use of telescopic sights, and would volunteer for duty.

In the instructional phase we were fortunate. The Exec. had commanded a reconnaissance company during World War II, and was well versed in both the theory and the application of scouting and patrolling. While some of the prospective snipers were shooting, the others were being schooled. All the knowledge and experience the Exec. had was being fed to the future snipers as rapidly as time would permit. After a week of this dawn-to-dusk routine he talked with the men individually. As a result of these talks some were returned to their units due to their own preference, others because they weren't suited to be snipers. The men were organized into three squads, four teams to a squad, and two men to a team.[1]

In under a month, the sniper company had proved its worth in action, inflicting many casualties during an abortive Chinese assault on American defences. Colonel Martin concluded: 'In surveying our sniper platoon program we found that some of our theories worked, others were workable with minor modifications, and some just wouldn't hold water. As time went on some of the snipers came up with innovations of their own which were better than some of our carefully planned theories.'

The experiences of the 2nd/5th Marines were amplified by those of 3rd Battalion, 1st Marines. In an account written by Major Norman W. Hicks, a newly arrived battalion commander decided to look over the terrain in no man's land:

When the shell-scarred slopes became visible at first light, he placed his binoculars in the bunker opening and gazed

out. Ping! A sniper bullet smashed the binoculars to the
deck while blood welled up in the crease in his hand.

The battalion commander, fortunately, was only
scratched, but he reflected that it was a helluva situation
when the CO could not even take a look at the ground he
was defending without getting shot at. Right then and there
he decided that something had to be done about the enemy
sniper. Now was the time to bring in the pin-wheel boys –
the Marines who could keep every shot within the V ring at
five hundred yards.

Sending for the S-4 [staff officer], the colonel learned
that within the supply section there was an adequate number
of rifles and telescopic sights. The colonel next sent for an
experienced gunnery sergeant who had spent considerable
time firing with rifle teams. He told the gunny what he
wanted: he then sat back and waited. His expectations were
completely fulfilled.

The gunny visited each company to pick sniper candidates. He outlined his requirements to the company commander. He wanted riflemen who possessed the
characteristics of good infantrymen. But above all, he stressed
the need for patience. This trait is absolutely essential, for a
sniper must remain still and alert for long hours, waiting for
the enemy to show himself.

Each company sent a number of candidates, and the
gunny selected approximately six two-man teams per company. Next he picked his sniper range, an area in the
battalion rear. He got the necessary target supplies, issued
weapons and 'scopes to the five dozen Marines, who commenced setting up a training range.

Soon the range was ready, and the gunny began an
intensive three-week course on sniping. Fortunately there
was no shortage of either .30 or .50 calibre ammunition at
that time, so the students were able to practise until
proficiency was attained.

Each student trained not only with the .30 calibre M1 rifle (or the '03 Springfield, depending on his preference), but with the .50 calibre machine-guns, fired single shot. 'Scopes were mounted on the machine-gun and they proved effective for ranges up to and beyond 1200 yards.

Lying for hours with their rifles sighted up the draws, the sniper teams trained together daily, learning their new roles well. When they finished the gunny's sniper course, they were qualified snipers in every sense of the word, and their future performances readily proved it. Returning to their respective companies, the men occupied camouflaged bunkers, which protruded only a foot above the ground's surface. These were artfully concealed, and the occupants entered or left only during the hours of darkness. The enemy found them extremely difficult to spot.

At the time the snipers finished their special training, enemy artillery and mortars were daily peppering both the MLR [Main Line of Resistance] and the outposts. Enemy snipers seemed to be in control. Then the Marine sniper teams were sent out to the various outposts. To spur them on, a case of cold beer was awarded to the men of each outpost that got 12 kills within a week. All hands turned to in helping the rifle experts in spotting enemy snipers. The change in the situation was fantastic. 'In nothing flat there was no more sniping on our positions,' remembers the Bn CO.

Only a week after the sniper teams went into action, the division commander came to test their efficiency. Where only a week before men hardly dared to stick up their heads, the two-star general strode the entire length of Item Company's MLR, armed with nothing but his walking stick. At the end, he jumped down and exclaimed to the company commander, 'By God, Spike, it works!' His voice still filled with astonishment, he added, 'What we need are more snipers on this front.'[2]

Gunnery Sergeant Francis H. Killeen had served in the Marines during the Second World War, and during the late 1940s had filled a vacancy in the USMC Scout Sniper School. Killeen was shipped over to Korea with the 7th Marines, and after a period as runner for his platoon leader, he was provided with a M1903A4 Springfield sniper rifle fitted with the long 8-power Unertl telescopic sight. Killeen describes his introduction to sniping in the field:

> Our first real chance to use the sniper rifles came at Su Dong Ni where our battalion was rushed up to the front and thrown headlong into a fire fight against a Chinese regiment.
>
> We got into a ditch alongside the road, and I immediately figured our range would be 400 yards uphill. I looked through my 8-power scope and could spot an otherwise unseen line of riflemen firing down at us by seeing the 'fuzz' of their projectiles. Lieutenant Davis, our 60mm mortar man, was right beside me. I told him where the enemy fire was coming from and in less than a minute he had three mortars in action and the fire lifted. We attacked and took the positions.
>
> This was the first time I found my sniper gun to be more effective than my M1 rifle. Usually we were in close actions where the M1's rapid fire could be more important than the long-range capability, but this time we were looking over a big valley at ridges 500 to 600 yards across the way. I chose a rock on the far ridge and got my lieutenant to spot my strikes with his binoculars. In that way, I made sure my rifle was still shooting where I aimed.
>
> In the late afternoon, columns of enemy began moving into position for evening festivities. I got off a couple of rounds, but without a spotter I could not tell if I was making hits. I got a BAR [Browning Automatic Rifle] man to register his rifle on the same rock I had used for zeroing. When he had his sights right we tried some team shooting.
>
> When I located the enemy I fired tracer at him or them.

Although tracer is lighter than the AP we were using, the trajectory was close enough and the BAR man, who could not see the enemy soldiers with his iron sights, simply attempted to catch my tracer with his bullets. The idea was to hammer the enemy with a decent volume of fire in the hopes that if I missed, the BAR man would get him.

The technique was instantly popular, and I soon had a light machine-gun and two more BARs creating the biggest beaten zone I ever saw. My lieutenant, James Stemple, 3rd Platoon A–1–7, got more riflemen into the fray, and we had the enemy falling all along their wood line.

In the heat of the action, and with the obvious success we were enjoying, I forgot about the 'hit and run' rule. We have to remember that the other guys also have people who can shoot. A bullet about one click low reminded me, and I cleared out just as a few more came into where I had just been.[3]

While still fighting in the Su Dong Ni area, Killeen describes a lucky shot against the Chinese:

The Chinese used mostly concussion grenades and one slipped in that knocked the wind out of me. I had a dozen pieces of shrapnel in my field jacket and my leg. Most could have been removed with tweezers.

The fire fight was over as fast as it started, but I was lying there clutching my scoped rifle when someone pointed out a Chinese across the valley waving a red flag with other Red soldiers cheering him on. My last shot had been in close, so I had to adjust the scope for 600 yards. I got into a hasty sling and fired at the flag waver. That shot may still be flying around North Korea because no one paid any attention.

Someone ordered me to the aid station and I started away, but the guy waving the red flag really annoyed me. Odd hand, with no sling – I even forgot to pull the scope back into position – I let a round go at the flag man. Down

he went, right on his butt, and his Red Army cheering
section hit the dirt.

No one was more amazed than me, but our Exec. was
trooping the line and saw the shot: 'Damn fine shot, Killeen.'
So, being a modest man, I accepted all of the congratulations
as if that kind of shot was an everyday thing.[4]

A platoon sergeant in the 3rd/5th Marines, John E. Boitnott was
a distinguished marksman who had trained as a sniper. Deployed
in a static sector, Boitnott and his fellow Marine snipers found
themselves increasingly called upon to counter the activities of
persistent North Korean marksmen. Operating with PFC Henry
Friday, who acted as a spotter, Boitnott was soon in action:

> Our position was outpost Yorke, about two miles forward of
> our front line and due north of the place where the peace
> talks were taking place. We were on the northern parapet of
> our hill, watching the valley for hostile movement. Friday
> suggested it was his turn for breakfast and asked permission
> to go through the centre trench to the reverse slope for chow.
> The centre trench ran sort of perpendicular to the front and
> had to be rushed through before an enemy took a crack at
> you.
>
> I was lying along the forward parapet and when Friday
> ran through someone took a shot at him. I thought I saw
> movement on a hill mass across the valley. The area was
> about 670 yards away. I called to Friday and told him to
> come to me. As Friday cleared the trench the enemy sniper
> rose again to fire, and this time I saw him clearly. One shot,
> one dead North Korean sniper. Lt Johnson verified the kill
> and reported it to the company commander.
>
> Friday and I teamed up, with him running the trench and
> me shooting the enemy for two more kills, which seemed to
> entertain about everybody on our side. Shortly thereafter
> word of our 'operation' reached someone less imaginative and

we had to knock it off. Over a two-day span, I made nine confirmed kills in nine shots from 670 to 1250 yards.[5]

Corporal Chet Hamilton was a sniper in the US Army's 7th Division, holding the line at the notorious Pork Chop Hill. Stationed at a strong point called Erie, Hamilton helped support an American assault on Chinese Communist positions. Hamilton recalls that the attack soon lost momentum:

I felt helpless watching from the sandbagged position on Erie until I noticed something. It was only about four hundred yards across the valley from the Chinese lines. My position put me on almost the same level with the Chink defenders on the other hill. In order for the Chicoms to see our troops and fire at them down through their wire as the GIs charged up the hill, they had to lean up and out over their trenches, exposing wide patches of their quilted hides. That was all I needed.

It had become a clear morning in spite of the smoke and dust boiling above the Chinese hill. The 4-power magnification of my scope made the Chinks leap right into my face. All I had to do was go down the trench line, settle the post-and-horizontal-line reticle on one target right after the other, and squeeze the trigger. It was a lot like going to a carnival and shooting those toy crows off the fence. Bap! The crow disappeared and you moved onto the next crow. By the time you got to the end of the fence, you got back to the beginning and the crows were all lined up again ready for you to start over. I don't know who the Chinese first sergeant was over there, but he kept throwing up another crow for me every minute or two. And I kept knocking them off the fence. The fight for the hill lasted about two hours. I was busy for the full time. The other guys on Erie came to watch, point out targets, and cheer when I zapped one.

The GIs never made it past the Chink wire. Heavy

Chinese mortar and artillery fire stopped the advance and
started cutting the infantry to pieces. The tanks scurried for
cover when the artillery fire singled them out; they weren't
much benefit on the steep grade anyhow. By the time the
GIs withdrew from the hill, dashing from rock to rock,
demoralized and defeated, my gun barrel was so hot that I
could hardly touch it. You could smell the cosmoline being
cooked out of the metal. I know I shot at least forty Chinks
before the attack bogged down and the enemy went back to
their burrows. Bodies had to be stacked up in the Chinese
trenches.[6]

Although the United States provided the majority of forces in
Korea, they were ably supported by other troops from the United
Nations, notably from Britain and the Commonwealth. Sergeant
Tom Nowell was a sniper with the 1st Battalion, the Duke of
Wellington's Regiment, winning the Military Medal while serv-
ing in Korea. Employing methods similar to those of Sergeant
Killeen in the US Marines, Nowell used his prowess as a
marksman to pinpoint targets for other, heavier weapons:

I was operating in a part of the Commonwealth front in
Korea just north of the famous river Imjin. From where I
was positioned on one late afternoon, in the dry weather, I
observed a party of Chinamen acting as a working party and
carrying supplies up the side of a feature code-named
'Pheasant', thinking that they were out of sight of our forces
and would be unobserved. The materials looked to be
ammunition and petrol, or diesel. They were obviously
stocking up for another go at our defences.

   There were about 10 or a dozen involved in this
operation, and if they were to continue their work uninter-
rupted they could have had a sizable amount of stores
brought up to the upper slopes by night-fall. I decided to
enlist the efforts of the local tank and mortar forces in our
area. They wanted to know the location and map reference to

this point, but I wasn't able to give them a good reference as I had not a map with me. We did, however, carry one or two tracer bullets that we used for target identification. It was arranged that I would fire a tracer into the area of the target and let the gunner OP pick out the grid reference.

The target was more than a thousand yards from where I was, but we were used to firing at those sorts of ranges. If I got a shot anywhere near to the working party, then the gunner could encompass that little lot in their ranging and consolidated fire power.

On the word, I let a tracer go at the centre of activity and I observed the tracer going across the intervening valley and on to the target. The shot struck one of the people carrying petrol and it ignited. It spilled out into a ball of flame and the luckless Chinaman fell to the ground and went rolling down the hillside. His comrades downed tools and went after him, grabbing him by the legs and trying to put out his burning clothes by throwing dirt and anything else they had to hand. It was a pure lucky shot and the ball of flame that erupted served as a fine target pinpointer for the gunners.[7]

The war in Korea saw the first practical experiments in developing .50 calibre shooting. For the most part, this involved fitting a telescopic sight on the M2 Browning machine-gun, which could be fired in single-shot mode, although adaptations of other weapons were attempted. Ralph T. Walker, a noted US gunsmith, experimented with old Boys anti-tank rifles while on assignment as a small-arms expert to the Chinese Nationalist Army in Formosa (Taiwan).

While fighting was under way in Korea, the Nationalists conducted raids on the Chinese Communist mainland, which inevitably brought retaliation from the Communists in the form of heavy artillery bombardments and the threat of invasion. Against this background, Walker recounts how the .50 calibre rifle came into being:

Just off Quemoy are two islands, one Nationalist, the other Communist. These were the outposts, with the Commies' artillery holding the largest of the two. The Communists could not take the small island as Nationalist artillery on Quemoy protected it. With a shallow stretch of water separating the two, it was too deep to walk but too shallow for boats. The 1100 yards between the islands made firing with a .30 calibre rifle a joke, due primarily to the incessant wind blowing across the water.

With American guns pouring into the arsenal at Formosa, we were pressed for storage space. While inspecting the warehouses, we came across some .55 calibre British Boys anti-tank rifles, undoubtedly used against the Japanese, but not one round of ammunition. Remembering the two islands, the idea to build a supreme sniping rifle using the big [M2] .50 calibre machine-gun round to buck the wind struck my fancy. With a small machine shop and the help of Chinese ordnance personnel we built the rifle.[8]

Walker and his team adapted the Boys to take the .50 calibre round and fitted a 20-power spotting scope (originally for tank use) using a homemade mount. Walker recounts how the Boys was used in action:

The Nationalist Chinese became interested in the rifles and we wound up building twelve of the Boys anti-tank guns into .50 calibre snipers. Picking the biggest [the Boys had ferocious recoil] and the best shots, they put them through a training program with the .50s. While they couldn't equal the accuracy of the original test rifle, groups ran about two feet long — still adequate for what we had in mind.

About half of the rifles and their new owners were sent over to the small offshore island. Although the front office would not let me accompany them, I read the subsequent report. On a usual basis the Chinese Reds would go down to the beach and bathe. At the first sound of any artillery firing at them from Quemoy, the Reds would scurry for cover and beat the first rounds coming in. With initial use of the .50 calibre rifles, their report barely audible, a dozen or so Chi-Com soldiers had their beach privileges permanently rescinded.[9]

After Korea the next major conflict involving East and West was the war in Vietnam. A nationalist struggle against the French colonial regime in Indochina was transformed by United States involvement. This culminated in a sustained military commitment by US ground forces between 1965 and 1972, in an attempt to save South Vietnam from Communist take-over.

Sniping had been allowed to decline in the US armed forces during the years between Korea and Vietnam, and the reintroduction of sniping was an often slow and painful business. The US Marines were to the fore in developing good sniping practice, although many company and battalion commanders did not understand how to use their sniper asset to best effect. This failing was highlighted by Master Sergeant William D. 'Bill' Abbot, a crack shot at Olympic Championship level. He explains how he set about rectifying the situation, instigating his own, highly productive sniper policy:

When I arrived at Camp Carroll in the Republic of South Vietnam on Christmas Day 1966, I was assigned the duty of platoon commander of the 3rd Marine Scout Snipers. I was instructed to report at once to the regimental commander, Colonel John Lanigan.

Colonel Lanigan informed me that, 'The snipers are scattered from hell to breakfast.' He said that they were not being used as snipers and that I would round them up, and retrain and refit them for duty as scout snipers. He said he wanted a unit he could call on to plug holes as needed. He was right. I found snipers being used on mess duty, guard duty, burning heads [latrine waste], and any other dirty task that needed to be done.

One of the company commanders that had a two-man sniper team assigned to him told me, 'I don't give a God damn who you are, you don't get these men until they are off mess duty.' I asked to use his field phone, and he said, 'Go ahead, call anybody you want, but you ain't getting them.' Colonel Lanigan requested that the company commander take the phone. I told the captain that 'Starboard 6' [Lanigan's call-sign] wanted to talk to him and he turned white.

The company commander's conversation consisted of a series of 'yes sirs' and 'It won't happen again, sirs'. When he got off the phone he told his first sergeant to get the snipers to regimental HQ on the double. He also turned to me and said, 'I owe you one.' He never collected.

While the scattered snipers were being assembled I had my platoon sergeant and squad leaders put together a training program that included refreshers in all aspects of scout sniper duties. I included dawn till dusk shooting, with a minimum of six hours shooting per day. The training included a lot of hand-to-hand combat, physical exercise, use of radios for artillery and air strikes, map reading, and other assorted stuff you would expect.

With the training under way I took a squad leader, rations for fifteen days, maps of our entire AO [Area of Operations], and went on a scout mission from Camp Carroll east to the ocean. We moved at night and slept little. We studied terrain during the day. When we reached the ocean north of Dong Ha we crossed the Ben Hai river and went west in the DMZ to Laos. We crossed the river to the south of the DMZ and returned to Camp Carroll.

There were several reasons for this trip: 1) to learn the AO we would be working in; 2) to check the NVA routes south; 3) to find the best locations to set up interdiction teams; 4) the former platoon commander would not go out on patrol and had failed to check his men assigned to the infantry companies. I did not make the same mistakes. To prove to my troops that I would go and that I had been there, I took plastic playing cards and hid them in fairly easy to find locations across the AO. I signed the cards with the date. A patrol was given the map location and a description of a card's concealment. When they found it they knew I had been there.

Our convoys along Route 9 from Dong Ha to Khe Sanh were sniped at regularly. To combat the annoyance we set up a system of two scout sniper teams in two jeeps with a squad leader and a radio operator who would work as a team. We put together ten of these teams and stationed them at appropriate points along the route. When a convoy was fired upon we responded from the east, west and Camp Carroll. We could have a team on the scene within thirty minutes anywhere along Route 9. We secured the route from so-called sniper fire.

I visited each battalion commander and every company commander to give them a run-down on how best to employ scout sniper teams. This duty turned out to be nearly full time due to combat injuries and the rotation of officers. The effort was worthwhile, however, and in most cases our snipers were used correctly and effectively.[10]

Not every sniper was fortunate enough to have a platoon commander of Abbot's resolve. In many cases it was up to the individual sniper to persuade his superiors, through combat success, that sniping was a worthwhile activity. In 1966, Thomas D. Ferran arrived in Vietnam, and after volunteering for sniper training (having been taught by Carlos Hathcock) he was assigned to M Company, 3rd Battalion, 7th Marines. He describes how his superiors initially had difficulty knowing what to do with him and his partner when they reported for duty at a hill-top base near Chu Lai:

When we arrived on the hill – which was surrounded by the NVA getting ready to go into the attack – we went to report to the First Sergeant, who said, 'Who are you guys?'

'I'm reporting into Mike 3/7,' I said, 'and I'm your sniper.'

'I don't have time for this,' he replied. 'Look, that's the skipper's rack, that's my rack, and that's your rack. You're our bodyguard.'

'Wait,' I said, 'I'm not going to be anyone's bodyguard. I'm a sniper, let me tell you what I can do for you.' In the end I sold them on the idea of sniping and they let me go to the weapons platoon. These guys received me very well; they had the mortars and the machine-guns and they knew how to use someone with a long-range rifle with a scope.[11]

Ferran – who eventually went on to amass a score of forty-one confirmed kills before being commissioned into the US Marines and subsequently transferring to the ARNG – realized that only results in the field would convince his fellow Marines of the usefulness of sniping. He describes how he made his first kill, which was fortunately observed by many Marines within their base:

Our first mission was on a hill south of Chu Lai in January 1967. We had just arrived and were getting ready to go out on patrol the next day, when I said to my partner, Pee-Wee, 'Let's sit in and see what's happening.' We saw this guy digging a hole about 1300 metres from us. He was clearly too far away for an infantry Marine to fire at. We surmised this guy was digging a spider hole; he was on a hill in a free-fire zone and he was digging a position that had clear zones of fire. We knew he was preparing this position to fire on a patrol some day.

I'd trained to fire out to 1100 metres, and when I graduated I had to be able to put a bullet into a ten-inch circle eight out of ten times at that range. I knew the VC was at about 1300 metres because it was two more football fields beyond the 1100 metres range we had trained on. At these long ranges the wind became a major problem. In the end I had to dead-reckon. I said to Pee-Wee, 'I'm going to aim this shot high and right, over this guy's head. I want you to spot for me.'

Meanwhile, unbeknownst to me, there was a group of Marines coming around to look at us. I was getting ready to fire my best shot. I had a perfect rifle and this was my first shot since graduating. I fired and missed. Pee-Wee had put the binoculars a quarter-turn out of focus so he could read the wind and follow the bullet trail going down to the target and see it impact. Pee-Wee called that I was low and to the left, about five and five. So I put the cross-hairs right there. When I fired that first shot and the bullet missed, the guy just looked up and couldn't believe he was being shot at. He took cover but resumed his duties within a few minutes.

When I hit him with my second shot, there was applause from the guys around me. We were elated because we were trying to sell sniping to the infantry, and here they could see that the concept of sniping could work. The VC body just lay there. We didn't go out to get it, as we were hoping that another VC would go out and try and retrieve it, then we would take him out also. But by then we were out on patrol with the infantry.[12]

Marine Steve Suttles arrived in Vietnam in January 1969, and, although a trained scout sniper, was initially assigned to the infantry as a basic foot soldier. Later he effected a transfer to a sniper role. His account details some of the problems faced by snipers from their superior officers, and how they were overcome with some spectacular shooting:

Snipers Parks, Jackson and I arrived at Hill 55 in late September or early October [1969]. Carlos Hathcock, the crack Marine sniper, had made history off this hill, and we later deduced that we had been sent to replace Carlos and others following Hathcock's heroic saving of burning Marines [which had caused Hathcock severe burns].

Hill 55 sits on the north side of a river facing Arizona Territory where fighting was unrelenting. To its north is a large mountain range slicing out of Laos and pointing towards Da Nang. Beyond the mountains is the Ah Shua Valley. Hill 55 with its Marine snipers was a large thorn in the side of our enemy.

We had a rear echelon type commanding officer who, it was rumoured, was assigned to the hill to punch his combat time ticket. The CO was a thorn to most of us because he was content merely to sit it out. A discouraging example of timidity occurred when a grunt patrol trying to come back from Arizona came under attack. We could see the action from out tower, but our request to provide support was repeatedly denied.

Finally Jackson and I with an M60 [machine-gun] crew and two others took it on ourselves to help out. We left the hill, and ran down across from a small island to the left of the trapped Marines. We waded out to the island where we laid fire into the surprised enemy and gave the Marines enough time to extract themselves from the ambush and safely across the river.

Of course we were threatened with court martial, which was so silly we did not worry. After the initial shouting we heard no more about the incident, but the CO hated me from then on.

It was shortly after the above incident that I recorded my long shot. While I was on watch in the tower, a pyjama-clad VC-looking type with a weapon walked along the opposite bank of the river. He did not know it, but he had just walked into our KD [Known Distance] range.

The snipers had recorded various targets and landmarks around the entire hill, and we had range cards showing locations and distances to many spots. The VC had walked to a fallen tree that we used for a target. The range to the tree was 1250 yards.

There was an officer on duty in the tower with Jackson and me. I asked for permission to engage and he looked at me like, 'yeah, right', but he OK'd the shot. I dropped the VC right over the log. The officer looked at me again and said something like, 'I don't believe it, but I saw it.' He signed my kill sheet. Of course I acted as if it was nothing special, but all the time knowing that as much luck as skill is involved in a shot like that.[13]

A sniper's first kill was always important to him, although there can be few more memorable than that of Corporal (later Sergeant) Jeffrey Clifford, a scout sniper in Charlie Company, 1/1:

The Hill 55 area was mostly flat with a few rolling hills. I guess a first kill is always an unforgettable experience. I certainly will never forget mine. 'She' was about eighteen and was carrying an M1 carbine. At the time I did not know that he was a 'she', and, of course, I did not know she was pregnant. The shot was overlooking a free-fire zone (which was a slightly elevated plateau) outside a deserted villa. A small VC patrol broke from the tree line and was re-entering it when I took out 'tail-end Charlie'. The shot was about 600 yards, through the lower back and out the front. The body and rifle were recovered, the kill confirmed. The grunts gave me credit for a 'two with one shot'. I took solace in staring at the sun.[14]

In the fighting to regain the city of Hue during the Tet Offensive of 1968, Clifford recorded an exceptional long-range kill, which he modestly put down to good fortune. The fact that such long-distance shooting was possible at all (Tom Ferran's first kill was at similar range, as was the shot by Steve Suttles) owed much to the quality of the Winchester hunting rifles used by the US Marines (later designated as the 7.62mm M40), but more to the exceptional shooting skills exhibited by American marksmen. US Marine sniper historians Roy and Norman Chandler computed that the forty-four men in the original 3rd Marine Division sniper unit had an average of 4.75 years per man in competitive shooting and a hunting experience of 8.3 years per man. As a consequence, Jeffrey Clifford's shot may have been as much a result of skill as of luck:

My 'long shot' (+1400 yards) was courtesy of Hue. Most would say it was not a probable accomplishment, and I admit that I had fired at least ten rounds at these same NVAs straight up the boulevard on the other side of the river without success. Finally I found a forlorn lieutenant belonging to a tank group and borrowed his range finder. No wonder I was missing, I was estimating 400 yards short.

The NVAs were moving gear in and around a large pagoda approximately twenty feet high. I put the cross-hairs several feet above the pagoda roof and fired. Regas (my ATL – Assistant Team Leader) could not see a strike so I told him to get some tracers from a nearby bull gun team. With Regas spotting I fired again. Perhaps it was the luck of the Irish, but we all (the Lt was still hanging around, looking through his binocs) observed the tracer's high arc and saw it enter an NVA just below his neckline. He went down as if the proverbial ton of bricks had fallen on him. The Lieutenant was excited enough to bring over a captain who looked through the scope to verify the heap. Ah well, chalk one up to luck.[15]

During the extended fighting for Hue, Corporal Clifford and his spotter Lance-Corporal Regas were involved in a dangerous incident:

Regas came in one day announcing that he had found a convent. It was a walled structure across from a wide boulevard which was 'Indian Territory'. The convent had a well and some food. We decided to shed our clothes and bathe. After splashing around like seals we put on new duds and decided we should earn our keep by shooting something. Up the stairs we went to the second floor and a nice room. I complimented Regas on a superior selection.

Regas believed every convent must have a store of wine and left to locate it. In the darkened room I set up about eight feet back opposite a large window. I was in 'shadow' with a pretty fair view of the boulevard and several buildings. I watched enemy tracers leave a second-floor window about three hundred yards down and to our right, evidently shooting at some Marines in a courtyard below. I took aim and waited for another enemy burst. Then I put a 7.62 [bullet] slightly above the muzzle flashes.

This apparently pissed someone off as a few minutes later I was on the receiving end of a B-40 rocket which came through the window and detonated against the ceiling (so I was told). The concussion blew out the floor, dumping fragment-riddled me, rifle and gear onto the ground floor, bleeding and covered with white plaster dust.

When I came to I could not hear, but a corpsman was working on me, and Regas (with typical Marine sympathy) was bent over laughing because I was again at least as filthy and a hell of a lot more foul than before our bath. That one I chalk up as my most notable.[16]

Among the top-scoring snipers in Vietnam – who include the US Marine Charles B. Mawhinny (103 confirmed kills) and the US Army sniper Adelbert F. Waldron (113 confirmed kills) – the most famous remains Carlos N. Hathcock. Not only did he accumulate 93 confirmed kills, he was a key figure in the development of US Marine sniping in Vietnam and in the post-Vietnam era. His exploits have become legendary among snipers. He holds the record for long-range sniping, shooting a VC weapons carrier at a range of 2500 yards using a scoped .50 M2 Browning. As winner of the United States Long-Range High-

Power Rifle Championship at Camp Perry in 1965 (awarded the Wimbledon Cup) his marksmanship was never in doubt, but he also proved supremely adept at transferring this skill to the hostile environment of Vietnam.

Hathcock's ability in field craft was formidable, as he demonstrated in a three-day stalk of an NVA general. In a recent interview Hathcock explained how he tracked down his quarry:

It was right at the end of my first tour in Vietnam, but I took the mission on myself because I figured I was maybe a little bit better than the rest of them. I was the one who trained them and I was supposed to be better. The stalk was about 1200 yards. It took me three days and nights – and the morning of the fourth day was when I completed the mission.

I came out of the tree line into open land, and I went on to my side. To go flat on my belly would have made a bigger slug trail, so I went on my side. Patrols were within arm's reach of me; I could have tripped the majority of them, but they didn't even know I was there. I was in their backyard and they didn't suspect a one-man attack. And I knew from the first time when they came waldergagging by me, that I had it made. This would be real good. So I just continued worming along.

There were two twin .51s [machine-guns] on my left, two twin .51s on my right. And I could see the NVA cooking their groceries, and I was wishing I was there to have a little bit of it – I was that hungry and thirsty. But you got a job to do, you couldn't let any of that enter into it – it was just you and your bubble.

I crawled over a little rise, with my escape route out to the tree line. I was in between all the .51s, which were set up for air attack, and I don't believe they could have got down that far [to engage me] – thank goodness! I saw all the guys running around that morning, and I dumped the bad

To run AT&T WorldNet Service on your IBM-compatible PC, you will need: 1) A 486SX processor or faster. 2) Windows®95, or Microsoft Windows® 3.1x, Windows® for Workgroups 3.11. 3) 8-16MB RAM and 15-66MB of available hard disk space, depending on platform (some additional temporary space may be required during installation). 4) A modem connected to a phone line running at 14.4 kilobits per second or faster, but not to a Local Area Network (LAN) or ISDN line.

To run AT&T WorldNet Service on your Macintosh,® you will need: 1) Macintosh® running System 7.1 or higher or Power Macintosh® running System 7.5.3 Rev 2 or higher. 2) Apple Open Transport 1.1+ (if using Open Transport). 3) 8MB RAM (or more for better performance) with Virtual Memory turned on, 13-20MB of available hard disk space, depending on platform. 4) A modem connected to a phone line running at 14.4 kilobits per second or faster, but not to a Local Area Network (LAN) or ISDN line.

☎ 1 800 WORLDNET, ext. 505
🖰 www.att.net/wns/endless

guy at 700 yards – the distance I had zeroed my rifle in and the distance on the map. I estimated the wind, the temperature, the humidity, the whole ball of wax – trying to run it through my mind real quick. And it worked out right fair.

Then I had to get away. When I made the shot, everybody ran in the opposite direction, because that's where the trees were. And it flashed through my mind, 'Hey, you might have something here'. So I went for that ditch, that little gully, and made it to the tree line. I about passed out when I stood up, but standing gave me a little bit better speed. I had to look out for booby traps and everything when going back to my pick-up point.[17]

Writing in the *Chicago Tribune*, the journalist Jim Spencer provided this addition to Hathcock's typically laconic account: 'Hundreds of ants that had crawled inside his uniform dined greedily on his sweaty flesh. Forced to relieve himself in his pants, he stank of urine. He didn't need to worry about that problem any more. All he had to eat and drink the past 72 hours were a few cupfuls of water, parcelled from his canteen. His mouth was dry. Blisters covered his arm, hip and knee, where he had rubbed them along the ground.'[18]

Not every sniper mission went according to plan; indeed a key test of a good sniper was his ability to adapt to rapidly changing conditions. Snipers obviously preferred to engage their targets at long range, safe from return fire by conventionally armed infantry. At times, however, a sniper might encounter the enemy head on. But if he kept his nerve the appropriate result could be gained, as this account from Tom Ferran reveals:

We were a six-man team (two-man sniper team plus a fire team backup) travelling with an infantry unit whose primary mission was to search and destroy, with a secondary mission to gather intelligence. We were the intelligence-gathering team. After a short rest break we quietly slipped away, crawling from our positions. This was to deceive any enemy scouts who may have been tracking us that the larger unit had deployed its recon element.

After travelling alongside a river, we took up a position of concealment waiting for the cover of darkness to help mask our movement into the village to obtain information on enemy activity. After a couple of hours, across the 30M river an NVA unit was staging to cross. They had been watching our infantry unit and now determined it was safe for them to resume their movement. As fate would have it, we were directly along the path of their patrol route. They were apparently anxious to get going as the Marine combat unit they had seen had obviously delayed their movement.

We were at 50 per cent alert. I was asleep. My team mate woke me and said to listen to the sounds. The sounds were that of the distinct metallic signature of AK-47 rifles. As we looked through the bushes, we saw some enemy movement across the river with one soldier getting ready to cross. He was obviously their point security, whose mission was to secure the far side of the river.

We all realized what needed to be done. Through hand signals we began establishing a hasty ambush. We allowed the NVA point security to come across and check out the area. Because of our good camouflage position he did not notice us. We were about ten metres from him. Our training taught us to wait for the optimum moment before springing an ambush. We needed the NVA point to think the area was secure and then summon the main body across — into our kill zone! After looking through his field-glasses he motioned his unit to cross the river. I recall my heart pounding so hard that my chest began to hurt. We needed to maintain our discipline and cool to make this ambush work with maximum results. The NVA began moving. We began our slow and deliberate movement into position. Just as the initial few NVA soldiers began arriving at the bank, one of the

Marines switched his M-16 selector from the 'safe' to the 'auto' mode. Doing so, the metallic sound betrayed us.

At this point the NVA knew what was happening. I had pointed my M40 sniper rifle in the direction of point security, thinking we still had a few moments before triggering the ambush. As I looked through the scope at this guy, all I could see was his eyes – the scope was on nine power! I began to tremble in the excitement at what was to happen. As a result, I could not hold the sights on target. I reached up to adjust the power in the scope to three power. As the magnification began to change I could see him turning in our direction with his finger on the trigger, ready to spray lead on our position.

I jerked the trigger of my M40 – a practice *not* taught in Marine Corps sniper school – and the ambush was triggered! The 173-grain match round ripped through his forehead. He was dead before he hit the ground. It was clear we achieved the total element of surprise, since there was no return fire by the enemy. The ten-second ambush seemed like for ever.

After taking out the point NVA, I realized very quickly that my M40, a highly accurate sniper rifle, was nothing more than a 'Civil War' shooting instrument in this situation. As our fire began to lessen, I laid my M40 down and ran to the NVA I had just shot. I wanted his AK-47 for my own personal protection. We began to pursue-by-fire the fleeing NVA on the other bank who were fortunate enough not to be at the wrong place at the wrong time. Someone yelled, 'Cease fire'.

With that we regrouped, called for our extraction and began shifting our position in the event we were counter-attacked. As we began searching through our spoils we counted twelve dead. There may have been others who were hit in the kill zone and were wounded and escaped, or were killed and floated downstream. Of even greater significance, however, were the documents we recovered, identifying the positions of fifty-seven of their recoil-less rifles and 82mm mortars. Also captured were the names of forty-two ARVN spies, and the identification of the troops we had ambushed – an NVA reconnaissance unit. The results of the capture of these documents triggered a battalion-sized operation.

After I returned to the 7th Marine Regiment on Hill 55, I debriefed the S-2/3 [staff officer] on the mission and proceeded to the NCO club. After a few beers I decided to inform the sergeant major, also at the club, on what had happened. I explained the ambush and that the M40 sniper rifle was useless in that type of situation. He agreed but began lecturing me in a very fatherly way, that snipers are not supposed to get themselves in those types of situation. I could not argue with a Marine who was in his third war and had clearly seen it all. Realizing this, I made a plea for a .45 cal pistol for personal protection (my AK-47 was turned in for intelligence purposes, as we were not authorized to use foreign weapons). As I began my sales pitch, he turned to me (after semi-ignoring me during this time), and said, 'Son, the day you need a .45 is the day you wished you had a rifle.' From that point on I carried two rifles, the M40 and the M16. He had convinced me!

Five years later, as a second lieutenant, I was checking into the 1st Marine Division at Camp Pendleton after my completion at The Basic School (TBS). I noticed an AK-47 with a plaque on the wall, among other type weapons captured during the Vietnam War. I went over to the AK-47 and began reading the inscription. Much to my complete surprise, the caption told of a six-man recon unit from the 7th Marines who ambushed an NVA recon unit. While reading this I began to reflect on the incident and the Marines who shared that incredible experience with me. I also couldn't help but wonder if I held the record for the shortest-distance sniper rifle kill in the Marine Corps.[19]

One of the stranger missions carried out by a Marine sniper was the downing of a Chinese officer operating with Communist forces by the Bon Son river junction south of Hill 65 on 4 November 1967. The sniper was Sergeant Bobby Sherrill of 26th Marines:

> At approximately 1100 hours the rain stopped, and the sun broke through an overcast sky. Zero wind conditions. Movement from my left: coming out of a distant tree line was a long skinny boat. An old man wearing a coolie hat propelled the boat with a long pole. A Chinese officer stood in the bow of the boat wearing a new-looking Chinese uniform, high stiff collar, puff hat, Sam Browne belt with pistol holster, also holding binoculars (lens gleaming in the sunlight).
>
> My rifle was an M1D with M84 scope built by 'Wild Bill' Johnson, RTE Barstow, California. I knew I had a one-shot try as the range was long and the boat had nearby cover. I was shooting from a bench rest with good sandbag support. I estimated the range at 850 yards. The M84 telescopic sight only adjusts to 650 yards, so I held the cross-hairs six to eight inches above the officer's left shoulder.
>
> I took a breath, let out some, squeezed carefully, and had good follow-through on the trigger. I got back on the M49 spotting scope in time to see the officer collapse over the side of the boat. The old man panicked. His hat flew away and he frantically poled the boat into the cover of the tree line.
>
> I immediately called a Sit. Rep. to higher command who relayed the report to G-2 Division. Somebody at Division wanted to call LBJ and the Pentagon to tell them that we shot a Chinese officer. They also wanted to send in a helicopter and a recon team to recover the body, but I knew the enemy had two M2 Brownings out there and they could trash the recon team.
>
> So I lied and said the shot was a probable hit. I did not want to see Marines die to advance careers in the G-2 section. I suspect that the Chinese officer was a rear-echelon pogue.

No one else would wear a parade uniform in a combat area.
A Chinese field officer would have been wearing camos and a
canvas shoulder holster. I know I made the shot and expect
it struck a blow to enemy morale.[20]

In contrast to the Marines' energetic attitude towards sniping in
Vietnam, the US Army response was slow and, initially at least,
ineffective. Whereas the Marines allowed their sniper teams to
work independently, the Army tacked them on to the back of
regular infantry patrols. As a result, they were prevented from
operating as snipers, reduced instead to acting as ordinary infantry,
albeit armed with a superior scoped rifle (in this case an 'accurized'
M14).

Eventually the Army realized its mistake and began to develop
an effective sniper policy. Much of the credit for this must go to
Major-General Julian Ewell of the 9th Infantry Division. A keen
devotee of operational analysis, Ewell was distressed to discover
the poor figures being gained by his snipers, and instructed his
officers to retrain and redeploy their men to achieve significant
results.

The ultimate success of the 9th Infantry Division sniper
programme – based around the requirements of good marksmanship and sniper-team independence – was such that it became the
basis for all Army sniper training in Vietnam. However, much
work had to be done to improve performance, as Colonel Robert
F. Bayard noted in a letter to Ewell: 'I am well aware that there
are few people in the Army today knowledgeable in engaging
targets beyond 300 metres, and that of this number there are
fewer still who have the ability of imparting their knowledge to
others.' Yet the work of the Divisional sniper trainers began to bear
fruit, as witnessed in this newspaper report of December 1968:

'Most of the kids we get in here are pretty green as far as
marksmanship is concerned,' said MSG Alfred B. Falcon,
NCOIC of the 9th Inf. Div.'s new sniper school. 'Spraying
the jungle with an M16 is a far cry from hitting a target at

over 700 metres with one bullet on the first try. It's a good program,' Falcon continued. 'The guys train on highly accurate match-grade M14 rifles and fire match-grade ammunition. All but two of the instructors are members of the "President's Hundred", the top 100 rifle shooters in the United States.'[21]

An interesting aspect of the Army's sniper policy was the widespread use of night vision aids. This report by Major Powell, commandant of the 9th Infantry Division sniper school, explains their use within overall sniper deployment:

Right now we are up to 148 kills and that has probably been upped since 1800 [hours] today. Most of the kills are coming at night with the Starlight Scope mounted on the M14; the VC aren't moving much in the daytime, but whenever one of our snipers gets off a shot they rarely miss.

The equipment is holding up real well, and we've had one-shot kills in daytime up to 800 metres and up to 550–600 metres at night. Most of the employment at night over here is with an ambush patrol. The snipers are given permission to engage one- to five-man groups and the rest of the ambush element holds their fire until needed. The snipers then maintain surveillance over anybody they knock down and pick off other VC trying to recover weapons, equipment, or the body.

Under the present set-up in the division, we will have a total of seventy-two snipers, six per battalion (equalling sixty), plus four per brigade (equalling twelve). The snipers are employed in pairs, with at least a five- to eight-man security element along. They are controlled and managed by the battalion commander and S-3, and are attached to the companies that have, or are likely to have, the most contacts. In addition to night ambushes, they have been used on berm, roads, bridges and firing sites as a security type mission.

We have been working in conjunction with the Xenon

searchlight with a pink filter. This is a very good combination when the natural light from the moon and stars is at low level [the Starlight Scope requiring a small level of ambient light to operate]. We also coordinate flares to help out when needed.

On daylight missions [the snipers] are being used in numerous ways, in both elements of cordon and searches placed at the back of villages to pick off fleeing VC when the village search is on, and in stay-behind elements when units are displacing to a new objective, to pick off VC following them.

The primary consideration that must be kept in mind when planning an operation here is to put the snipers in the position with good fields of observation and long fields of fire. The units that are taking a little extra time in the planning phase of an operation are the ones racking up kills with snipers.[22]

Of the snipers in the 9th Infantry Division, the most successful was Sergeant Adelbert F. Waldron, who achieved a spectacular score of 113 confirmed kills, many of them taking place in night actions (sometimes using a noise suppressor). The following after-action report details Waldron's success in night sniping:

Sergeant Waldron and his partner occupied a night ambush position with Company D, 3/60th Infantry on 4 February 1969 approximately three kilometres south of Ben Tre. The area selected for the ambush was at the end of a large rice paddy adjacent to a wooded area. Company D, 3/60th Infantry had conducted a MEDCAP and ICAP in a nearby hamlet during the day, hoping to gain information on Viet Cong movements in the area.

At approximately 2105 hours, five Viet Cong moved from the wooded area towards Sergeant Waldron's position and he took the first one in the group under fire, resulting in one Viet Cong killed. The remaining Viet Cong immediately dropped to the ground and did not move for several minutes. A short time later, the four Viet Cong stood up and began moving again, apparently not aware they were being fired upon from the rice paddy. Sergeant Waldron took the four Viet Cong under fire, resulting in four Viet Cong killed.

The next contact took place at 2345 hours, when four Viet Cong moved into the rice paddy from the left of Sergeant Waldron's ambush position. The Viet Cong were taken under fire by Sergeant Waldron, resulting in four Viet Cong killed. A total of nine enemy soldiers were killed during the night at an average range of 400 metres. Sergeant Waldron used a Starlight Scope and noise suppressor on his match-grade M14 rifle in obtaining these kills.[23]

The qualities required of a top sniper, such as Waldron or Hathcock, included mastery of the technical details of sniping – marksmanship, fieldcraft, tactics – and the right temperament. Psychopaths and other psychological casualties of the war were not welcomed by sniper trainers. During Hathcock's second tour of Vietnam, one of his sniper trainees suggested shooting some Vietnamese farmers: 'We let him go,' Hathcock said, 'put him back in the line company. He just wanted to shoot somebody – anybody. Snipers aren't trained that way. You have to train your mind to control every emotion you have.' Speaking later of the requirements needed for a good sniper, Hathcock observed:

I like someone who is quiet, malleable, knowledgeable, able-borne, very highly intelligent, with a knowledge of fieldcraft and weapons, and an ability to observe and get their observations down in writing. You've got to have the utmost patience; I've lain three days in one spot, just observing. To be a good shooter you need a good head on your shoulders, and to be able to absorb everything that is going on around you. You need a pure, absolute concentration on the job you are doing – I call it getting in my bubble.[24]

The need for patience was underlined in an account from Mark Limpic, an experienced Marine sniper:

> I remember how hard it was to train – to instil – patience in new snipers, the patience to sit on your butt for hours, waiting to get a good shot. Most new men get pretty excited upon seeing an enemy or two, and want immediately to get a shot off. One particular incident comes to mind.
>
> I took a couple of scout snipers out from Hill 65 and set in by the river. After a few hours several NVA soldiers came out of the undergrowth to cool off in the river. My snipers wanted to crank some rounds into them – so much for patience.
>
> I held them back, told them to sit tight, and to observe. I reasoned that more swimmers would probably come, and those who were already there would get deeper into the water – which is damned hard to run in! We waited, and they came. We got off a number of easy shots and some that were harder. We obviously screwed up their party. I cannot remember how many we killed, but there were more than a few. All due to our patience.
>
> That was one of the times I feel I taught the young studs to do better. They learned that patience is essential. Most Americans can't sit still for more that ten minutes, and the self-discipline to do so requires convincing training.[25]

Apart from these personal attributes, a sniper had to have a good relationship with his partner. If this relationship broke down then the mission would suffer. Tom Ferran provides this revealing account of sniper–partner relationships:

> The infantry will tell you that they operate with the closest bonds to each other – a band of brothers – closer even than

to their own wives. With snipers it's an even tighter bond because you're operating autonomously. And you must completely rely on just one other person.

There were some problems with one or two snipers who were not up to the high standard of professionalism that I thought was necessary. One was married and brought the 'baggage' of his wife and child with him on missions. He seemed a liability to me. We were there to kill the enemy, not talk about kids! That just didn't go over well with me; he was getting into my space and affecting my ability to concentrate. I was single and I had made up my mind that I was going to make a contribution to this war. I was very serious about this business, that's why I volunteered for the job.

About three weeks into our assignment we parted as team-mates. He didn't have the mentality to be a sniper; he had all the technical requirements but he didn't have the psychological requirement. In one incident during this first assignment, we were on patrol and had taken up a defensive position for the night. It was very foggy and dark, with the infantry spread out over 360 degrees, each man perhaps fifteen metres apart. My team-mate decided to leave me and check out where the infantry lines were, even though I'd told him not to leave. We could hear the enemy movement to our front.

He was supposed to have been gone for a few minutes at the most, but after forty-five minutes he still hadn't returned. The night was very eerie, and you just weren't sure what was going on. The elephant grass to my front started moving. I picked up my M14 and pointed in the direction of the movement. I called out, 'Halt!' There was nothing, no reply. I repeated this challenge two more times. I was sure that it was an enemy probing the lines and not a Marine. I started to pull back on the trigger, safety off and on full automatic. The trigger had a five to seven pound pull, and I knew I was at the four pounds pull point. All I had to do was blow on the trigger and it would go off.

The elephant grass continued to move when all of a sudden my partner came through the bushes and called out my nickname. Once I realized who it was I slowly released the trigger, and took the butt of the rifle and popped him in the face, almost knocking him out. I said, 'You son of a bitch, I just almost killed you!' After that close call I got a new partner.

It was good to change partners periodically so that you didn't get too comfortable with each other and as a result get careless. If you relied too much on each other you tended to relax, and as a result your guard was down. In the infantry you tended to rely on the other guy; it was part of their culture and spirit; it gave comfort, you were part of a family. As a sniper I did not like that. I liked to rely on myself and thus keep my edge.[26]

Despite the high level of training and competence displayed by US snipers in Vietnam, mistakes could occur. Ferran cites one small instance, which was nevertheless enough to make him determined not to slip up in the future:

One day I put my rifle against a tree forgetting to put the lens caps over the rifle scope. I left the rifle there for a good period of time. When I picked it up I didn't bother checking it before I started back on patrol. Stopping for a rest, I got into a good sniper hide and I looked through my scope. I noticed that the vernier scale had been destroyed. The sun had come over and gone through the scope, melting the rangefinder. Fortunately, it didn't damage the optics. This incident taught me to pay a little more attention to detail. The use of the vernier scale was not important to me as I didn't use it to determine distance. The idea, however, that I'd screwed up bothered me because you're supposed to know exactly what you're doing at all times. I then knew that we weren't infallible.[27]

The sniper was well placed to undermine enemy morale, as Ferran explains: 'The idea of having your head blown off before you even hear the report of the rifle shot will demoralize a military unit quicker than an artillery round coming in. They can react to artillery because they can hear it, but not to a sniper's bullet.'[28]

Marine sniper Craig Roberts further explains the special nature of the sniping war:

> We turned the tables on them; now, we were the enemy who hid in the jungles like ghosts, who killed unseen and then vanished into the shadows of the bamboo.
>
> Killing like this was personal, not like in a fire fight when it was difficult to tell who killed who. Through the scope you could see the expression on their faces in that instant before you sent a bullet screaming into their flesh. Mostly they were young faces, some no older than sixteen. I was just nineteen, so their youth had little impact. We were doing something personal to avenge Marines who had been killed and maimed by mines, booby traps and snipers.
>
> Going to the enemy on his own turf and killing him was better than waiting until he came on yours. We were seeing the enemy – and he was ours.[29]

Some of the complex moral and psychological consequences that a sniper has to face are contained in a short statement by Tom Ferran. He expresses feelings that have struck a chord with many other snipers:

> Sniping is a very personal experience. You look through your scope and you see what the person looks like. You realize they are human. You've got complete control over their life, you've become God-like, in that once you pull that trigger and take them out – you just owned their life. As a sniper you carry that 'last sight picture' with you for the rest of your life.[30]

Chapter Seven

Small Wars and the End of Empire

After the Second World War, Britain began the process of dismantling its empire. To ensure an orderly transfer of power to the former colonies in Asia and Africa, the British Army was involved in a number of small wars. The long conflict to prevent a communist take-over in Malaya – where thick jungle made sniping difficult – was succeeded by campaigns in the Middle East and Africa.

The British crown colony of Aden had exercised influence over southern Arabia since the nineteenth century. Political unrest in the area during the 1960s, coupled with the knowledge that Britain would eventually withdraw from Aden, led to armed conflict between rival nationalist groups eager to seize power. Not only did the groups fight among themselves but they also attacked the British.

The open, rugged terrain of southern Arabia provided good opportunities for long-range shooting. This was particularly so in the Radfan, a mountain range running inland from the port of Aden. British patrols advanced deep into the mountains to tie down and catch the elusive rebel tribesmen. The Special Air Service (SAS) provided reconnaissance to the main British force, but, as this account from April 1964 explains, they were also involved in sniping operations:

> I was laying up in an OP [Observation Post] overlooking a
> deep wadi, up in the mountains of the Radfan near the

Yemeni border. It was hell on earth in those temperatures. The sunset brought some relief and let me go to a well nearby to draw water. On the way down I spotted two guerrillas who obviously had the same idea. I got into position behind loose rocks and brought them into my sights. They were about 400 yards away and night was falling fast. The lead man went down with a round in the head, and his mate didn't even have time to raise his weapon before I'd dropped him too.[1]

By 1967 the fighting had moved into the city of Aden. There, the British were increasingly beleaguered by the rebels who specialized in urban terrorism, throwing bombs and grenades and sniping amidst a warren of streets. Guardsman Mike James was stationed at the Detention Centre at Al Mansoura, and was a member of the infantry company guarding the Centre from outside attack. He recounts how he had to endure constant sniping:

Over a period of time we were receiving casualties at the GPMG [General Purpose Machine-Gun] position, some fatal. Rumour had it that there were some mercenaries from Eastern Europe in the pay of one of the political parties. These mercenaries were being credited with the sniping. I never discovered if this was true or not; however, I can reveal that whoever the sniper was, he was very good.

One incident I witnessed was when the main gate of the Detention Centre was opened to allow a Ferret armoured car to go on patrol in the area. Warning was given every time the gate was opened for all to keep well clear because of the incoming fire from the houses around Al Mansoura. The Ferret had not even cleared the gate before its commander was shot between the eyes. This shot had cleared the inch gap between the armoured plating in front of his command position. The driver immediately reversed back inside and I and a company sergeant major removed the lance corporal from his turret. He was rushed to the medical centre but unfortunately died soon after.

Sniping was a regular occurrence throughout Aden. Many a time, when returning to Al Mansoura from Khormaksar Causeway or Steamer Point, we would be marooned at the Al Mansoura roundabout because of heavy sniper fire, and the only way of returning to the Detention Centre was by armoured vehicle. The only time we had a brief break from sniping was when the Arab–Israeli Six-Day War was on. We actually waved to the volunteers as they were going up country to the Yemen to join the Egyptians. They in turn waved back. Some eight days later they returned and the sniping continued.

One particular incident will remain with me for the rest of my life. At about 0600 hours I went up to one of the GPMG positions to speak to the sentry, who happened to be from my own Household Division, an Irish Guardsman. We were just looking over the walls at Al Mansoura and the surrounding areas, watching the sun rise. I was leaning on the sandbags with my head resting on my right hand. After a few moments in this position I said to the Guardsman, at the same time moving slightly to the upright position from where I was leaning: 'I feel that someone has got a bead on me.' I did not complete the sentence before a shot was received through the sandbag on which I was leaning. Tracing the trajectory of the shot, I realized that I would have been shot in my right temple had I not moved. Prior to going to that position, having been on duty all night, I was feeling tired. Suffice to say, after that incident I soon woke up and realized how stupid I'd been by allowing myself to become a target.[2]

The Arab rebels did not have it all their own way. The British fought back with their own snipers, most of them coming from the Royal Marines. They despatched marksmen to hold positions high up on the ring of crags that surrounded the Crater district, the scene of much terrorist activity. One of these snipers was Marine Mick Harrison, a former poacher and a crack shot who first gained sniping experience against the EOKA guerrillas in Cyprus. In this account he describes his mode of operation:

> The terrorists in Aden had been shooting up the Argyles or the Northumberland Fusiliers and they wanted snipers, so I was called for. I got in the back of a Land Rover, and they had got my kit, my rifle and scope. I checked my ammunition — Vickers .303 7Z rounds. They told me to go up the mountain, and for fourteen hours a day, for four days, I made my base in an old, ruined Turkish fort. Each morning I hauled myself up with a rope while it was still dark, around 5.30 or 6.00. When I got up there I hid. The important thing wasn't to go pooping off at anything. I saw I was all right for water before I went up, and in that climate you needed plenty. It was just a question of using training. Just a matter, so to speak, of giving the other man enough rope to hang himself.
>
> I had to make the terrorists show themselves. What I did was to expose myself to them deliberately. I'd kill one, then I'd move to a new position. They were about four to five hundred yards away. I'd get up and wave to them to draw their attention. It was the old idea of bringing your quarry to you. They thought they were good and I let them come to me. The first person I shot came out all dressed in black, and I remember I shot him in the throat. They wanted to go to Allah's garden, and I just paved the way for them.[3]

Marine Harrison was credited with killing or wounding eight terrorists for the expenditure of eighteen rounds of ammunition, and was specially commended for gallant and distinguished conduct. His citation read: 'He was in a position overlooking the Crater under sporadic small-arms fire and was the target of a blindicide [anti-tank grenade] attack. Over a period of four days, working with another sniper position, he systematically eliminated terrorist snipers opposing him so that all terrorist fire ceased during daylight hours.' On a subsequent occasion, Harrison was equally effective with his rifle:

> One Friday I was looking through a spotting scope and I saw a group of Arabs on this roof top. I thought there is something going on there; instead of praying like they were supposed to, they were arguing. In Aden at the time there was this high ranking officer – a general from the Irish Hussars, the Irish Guards, or something – and sure enough the Arabs ambushed the armoured car he was travelling in. I had to ask permission to open fire, and while I was waiting on the radio, my boss man came on line and said, 'Harrison, get on with it.' I opened fire and got three of them.
>
> I didn't bother with any back-up; it was just me against them. The sniper is the loneliest bloke in the world, and that's how I liked it. When I came back, the others could smell it on you, and they all wanted to get away. You didn't have any friends.[4]

One of the last of Britain's small wars involved retaking the Falkland Islands following the Argentinian invasion in 1982. Despite the successes gained in conflicts such as Aden, sniping continued to languish in the British Army. For those battalions rushed south in the Task Force, sniper training had to be hurriedly improvised. Of the two paratroop battalions despatched to the Falklands, 2 Para was fortunate in having the services of David Cooper. The battalion padre, Cooper was also a shooting enthusiast who trained the battalion snipers.

While sailing south on SS *Canberra*, David Cooper continued to run various shooting courses, as well as zeroing-in some newly acquired night sights. The daily release of the ship's rubbish enabled paratroop snipers to hone their skills on a moving, if not live, target. The lack of official interest in sniping forced those on the spot to do the best they could, as Cooper explains:

Because of the shortage of rifles the sniper pair was – on the whole – matched on their common zero. Everybody would shoot with a specific rifle, and from their shot groupings we could then identify which individuals had much the same zero. And as long as they could get on together, they then became the pair. The battalion was issued with eight rifles which made eight pairs, but we also took our own target rifles – six in all – which the battalion had bought.

A problem we had on the Falklands – as everybody did – was the amount of weight that had to be carried. Not least of all because the snipers wanted sniper-quality ammunition, and the re-supply was almost certainly not going to be able to get sniper-quality ammunition to us. So the snipers had to take as much ammunition as they could carry.

I managed to persuade the mortars to part with their laser range finders. They were heavy but effective, although at Goose Green they did tend to attract fire, but that tended to be more the result of a sniper not being particularly discreet over where he was, sticking his head over a gorse hedge.[5]

Much of the fighting on the Falklands (especially the set-piece battles) was conducted under cover of darkness, making night sights particularly useful, although there were limitations in a sniping context. In one of the key encounters of the war, a large force of well-armed Argentinians was attacked by the numerically inferior 2 Para. The commander of 2 Para, Lieutenant-Colonel H. Jones (known to his colleagues as H., and posthumously awarded the Victoria Cross), was determined to take Goose Green with the utmost speed and force. As ever, the correct deployment of snipers was crucial, as Cooper relates:

We used night sights at Goose Green, but it was there that I had the one real disagreement with H. – and he was wrong, and I am sure he would have admitted it. At the start of the battle, he put anti-tank weapons and snipers on a promontory around a loch, about a thousand metres from Boca House. It was a long shoot at extreme range, but he wanted to put fire on Boca House ruins should it be needed. I said to him that it was too far for snipers, especially as we were firing over water and there would not be a lot of indication of wind strength, and it could be very windy. The fire would be totally ineffective. We were shooting at night at targets that couldn't be clearly seen. Even with a night sight we had absolutely no indication of strike; you hit the target and it falls over, but you can't see where you missed and make corrections.

The bigger problem was that the snipers would take a hell of time to catch up. And this was exactly what happened. [Major] John Crossland [commander of B Company] got pinned down at Boca House, and the Milans [anti-tank missiles] and snipers couldn't do anything, and there was a three- or four-hour wait while they tabbed round to catch up with everybody, by which time H. was dead and the situation had changed radically.[6]

Despite the death of their commander, 2 Para reorganized and steadily began to press the Argentinians back to their defences around the settlement of Goose Green. As padre, Cooper was not directly involved with the sniping, but was responsible for casualty clearance and other matters at battalion headquarters. But his particular skills were still occasionally called upon:

I did one or two stalks where I did nothing more than call the wind. I wasn't necessarily a better technical shot, but I could call the wind better than anyone else.

The snipers tried to help out where companies were being held up. On one occasion we had to move quickly when we came under anti-aircraft fire. We'd obviously been seen and had to move back into dead ground fairly nimbly. Although the ground was open, it did give us quite a lot of scope for movement in dead ground, until you were within range. Then it was a question of suppressing fire, while the companies or platoons moved. That amounted to putting down harassing fire into the Argentine bunkers.

In one incident when I called the wind, the sniper fired and a white flag appeared out of the bunker, which was a good 600–700 metres away. On the shooting range the signal for a miss is a white and red flag, and the sniper made the comment that they were signalling a wash-out. Clearly, the Argentines were unhappy about the rounds going through the slots in their bunkers, but there was not a lot we could do when someone 700 yards away wants to surrender.[7]

After the British victory at Goose Green, the Paras rejoined the main force advancing towards the Falklands Island capital of Port Stanley. The Royal Marines formed a substantial portion of the British invasion force. Corporal Steve Newland from 42 Commando was awarded the Military Medal for his contribution to the assault on Mount Harriet, one of the Argentinian positions outside Port Stanley. The Royal Marines were constantly engaged by Argentinian snipers using night sights. Corporal Newland, working closely with Corporal 'Sharkie' Ward, describes his response to the problem:

All the time we were lying there rounds were ricocheting off the rocks at us and the cold was freezing our bollocks off. On the radio I heard Sharkie talking to his boss. He said, 'We're pinned down by a sniper and we can't move.' I thought, 'Right, someone's got to go for this bastard.' So I took off my 66 shells [for the 66mm light anti-tank weapon], got on the radio to our boss, and said, 'Wait there and I'll see what I can do.'

I crawled around this mega-size boulder, rolled into cover and looked around the corner of this rock, thinking that the sniper had to be there somewhere. There was more than a sniper – there was half a troop! About ten of them were lying on a nice flat, table-top rock, overlooking our positions. It was perfect for them. They had a machine-gun on the left and the rest of them were lined out with rifles. Every time one of ours tried to move forward, one of them would shoot at him, so it looked to us as if there was only one sniper who was keeping on the move. They were waiting for us to break cover and try and clear this one sniper – then they would just waste us with their machine-gun.

I sat back behind this rock and whispered down my throat mike to Sharkie about what I'd found. I picked up my SLR [Self-Loading Rifle], changed the magazine and put a fresh one on and slipped the safety catch. I then looped the pin of one grenade on to one finger of my left hand and did the same with another. I was ready.

I pulled one grenade, whack – straight into the machine-gun. Pulled the other, whack – straight at the Spics. I dodged back around the rock and heard the two bangs. As soon as they'd gone off I went in and everything that moved got three rounds. I don't know how many I shot, but they got a whole mag. I went back round the corner of the rock, changed the mag and I was about to go back and sort out anyone who was left, when Sharkie called on the net: 'Get out! We're putting two 66s in.'

The 66s exploded and the next thing I heard was Sharkie on the radio again: 'It's clear. They've given up. Go back to where you were and make sure they don't get out the back.' I went up by a different route and as I rounded this rock, I saw one of the guys that I'd hit. I'd only got him in the shoulder but he'd gone down like the rest of them, and in the dark I'd automatically thought he was dead. But he was far from that, because as I came back round the corner he just squeezed off a burst from his automatic. He must have realized he was going to die unless he got me first. I felt the bullets go into both my legs. I thought, 'Shit, the fucker's got me.' I was so angry, I fired fifteen rounds into his head.[8]

Three days after 42 Commando had secured Mount Harriet, 2 Para was ordered to assault Wireless Ridge, the last obstacle in the way of Port Stanley. Although the action was a success, the new battalion commander had decided not to use the snipers at battalion level, as David Cooper recalls:

> At Wireless Ridge our snipers were less effective when they were simply parcelled out to the companies, as company commanders didn't have any great awareness of using the snipers as a weapon system. The snipers tended to be told to go along to beef up a platoon that was a bit short on men. And, of course, it wasn't much fun being a rifleman with a bolt-action 10-round magazine rifle, when everyone else had an automatic or semi-automatic weapon. The sniper's great advantage of concealment was being thrown away when he was simply advancing as another infantryman with an ineffective rifle for the task in hand. And goodness knows what he was expected to do when he finally closed with the enemy because he hadn't got a bayonet either. But this was the situation in the Army for many years.[9]

The swiftness with which the war in the Falklands was concluded contrasted markedly with the long, drawn-out conflict in Northern Ireland. Originally, the British Army had been despatched to the Province in 1968 to protect the Catholic population. Soon, however, the Army found itself supporting the civil authorities against extremist terrorists, of whom the IRA was the most prominent. Patrolling the streets of Belfast or Londonderry, the security forces were vulnerable to terrorist gunmen. Many of these were merely IRA 'cowboys' taking pot shots at the British, but others were more competent. A sergeant in the Royal Green Jackets noted the activities of one such terrorist:

> We had a guy killed and two others injured by a single shot through the back window of a Saracen [armoured vehicle]. The round passed through the head of the first man, killing

him outright, took out an eye and damaged the nose of the
second bloke, and ended up in the third guy's ass. The
gunman had lain down on the table in the back kitchen of a
house they'd taken over, propped up the letter box with a
pencil and fired through the whole length of the house, out
of the door and into the back of the passing vehicle. It was
either a very lucky shot, or the gunman was very good. The
Belfast IRA had a guy we nicknamed 'One-shot Willy'
working for them at the time, and we always reckoned he
might have done it.[10]

The tactics used by IRA gunmen to secure good vantage points to
fire on the security forces is described in this report by Belfast
journalist David Barzilay:

On 16 May at about 1415 [hours] two armed terrorists
forced entry into 131 Castle Street. One of them held the
owners, two elderly sisters, hostage whilst the second went
up to the second floor of the building to prepare a snipe
position. He knocked a 6-inch by 3-inch hole in the wall of
the bedroom at floor level which gave him an excellent fire
position and a view of the Castle Street Segment gate. At
1520 one high velocity shot was fired at the sentry Pte 'Ty'
Lawrence of City Centre Company who was standing behind
the blast wall. The round ricocheted off a wall three feet
from Pte Lawrence's head and clipped the bridge of his nose.
A follow-up was mounted immediately but the terrorists
were able to effect an escape as the firing point was initially
extremely difficult to locate.[11]

An officer in the 1st Royal Anglians, undertaking a two-year tour
as one of the resident battalions, describes the effect of a sniper
casualty on his own battalion:

I think the one that will stick in my mind most of all was a
soldier I was talking to on the radio. It must have been a

chance in a million snipe. He was sitting on his radio in a sandbag emplacement with a little slit for him to look out of, a form of observation post, when a round came straight through the slit and he was hit in the head. I remember him actually saying on the radio mid-sentence, 'I've been shot', although I'm convinced he was clinically dead. It just went dead after that. I reported it and said, 'Someone's been shot. I can't raise him,' and then they got on the radio and reported that they'd had a contact. That was the first one we lost. It's very difficult to describe your feelings – anger, frustration, shock. I think it was the fact that he was a popular lad and a married man that brought it home, having to explain to his wife and seeing her face, seeing her leave the Province with the children. That was the difficult part of accepting it.[12]

Rising casualties from IRA gunmen forced the Army to take protective measures. New defensive tactics were introduced to make patrols less vulnerable, but the problem remained of returning fire against terrorists. As Northern Ireland was part of the United Kingdom, soldiers had to operate within the many constraints of British law; not only could any fatal shooting cause mass uproar but a murder charge might be laid against the soldier concerned. A carefully controlled response was essential, and in such an environment, the accuracy of a sniper could be invaluable. Some of the moral and legal aspects of sniping in Northern Ireland are explained by a British sniper officer:

We had a job to do. You have a moral superiority and have to believe that what you are doing is fully justified and is lawful: what you are trying to do is eliminate a terrorist. In this way your very first shot is your most difficult one. You've decided that this is a legitimate target, and it's that moment of squeezing the trigger. In the back of your mind, particularly if you are an officer, you can see the courts martial, the boards of inquiry and all those other things. It's

also what goes through the mind of many soldiers. Because
we operated in three-man teams – with one man looking
through a scout regiment telescope, which is a ×20 magnification – we were able to see and agree that the target was a
legitimate one. We were looking at armed targets who were
either threatening us or those it was our duty to protect,
which under the rules of engagement were legitimate
targets.[13]

As the IRA grew increasingly active during the early 1970s –
engaging in a direct shooting war with the security forces – the
need for snipers became increasingly apparent. One Royal Marine
from 40 Commando remembers how he was recruited:

> Prior to our '72 tour [of Northern Ireland] it was realized
> that none of the trained snipers in our unit, all being NCOs,
> would be available for that role. The best shots from each
> troop from the fighting and support companies were therefore
> asked to volunteer as snipers. I was the obvious choice in my
> troop and I joined eleven other Marines for six weeks training
> under an efficient and seasoned sniper instructor.
>
> I think I was probably the only one in the twelve to have
> previous service in the Province, and one other Marine had
> considerable experience of sniping in Aden six years pre-
> viously. We knew well in advance that our tour was to cover
> the area of Belfast city centre, including the nationalist areas
> of New Lodge (which by this time was a 'no-go' area), Docks
> and Bawnmore, and the loyalist areas of Fort William,
> Skegoneill and Rathcool. The final part of the unit training
> took place at Lydd and Hythe training areas, where as well
> as ample live firing ranges, an old married quarters estate
> had been turned into an 'IRA enclave'. The sniper section
> acted as 'enemy' for the rest of the unit, setting up sniper
> shoots and bombings, sometimes running rings around them,
> sometimes getting caught and interrogated.
>
> The .303 No 4 (T) sniper rifle was in the process of being

replaced and we were issued with brand-new 7.62mm L42s.
There were no brackets to fit the IWS 'Starlight Scope'
[night sight] to the L42, so we all also had an SLR with IWS
for night work. This meant that when travelling to N.I. we
had far more kit to carry than everyone else, and as we went
to this war on public transport every move was a nightmare.
We had to change trains and platforms at Birmingham, and
our first action of the tour involved a drunken Irishman who
decided to assault us with verbal abuse, which didn't last
long – my troop sergeant saw to that![14]

Once in Northern Ireland, British snipers were assigned a variety
of tasks, more often acting in a reconnaissance role. One British
Army officer, in charge of his battalion's sniper section, wrote this
report following an incident in Coalisland in 1972:

On the 29th January at 0430 hours, a three-man OP
[Observation Post] consisting of myself, Lance-Corporal
O'Riley and Private Smith [both names changed] was placed
in position on the roof of Stewart's Mill in Coalisland. Our
task was twofold. Firstly, to act as an OP for the SP Coy
[Support Company] and, secondly, to act as an anti-sniper
post in support of the planned operation for the dispersal of
the projected Coalisland–Dungannon march, scheduled to
take place at approx. 1300 hours on the 29th. Due to the
planned position of the barriers it was accepted that our OP
would be behind the crowd, and we would be cut off from
our own troops. The OP was to be covert until 1300 hours
and then when the crowd formed, become deliberately overt
to discourage anyone in the rear of the crowd attempting to
snipe at the troops on the barriers. The IRA had a habit of
parting crowds, shooting through them and then closing
them up again.

From 0430 until 1300 we remained hidden. The only
way this could be achieved was lying perfectly still in a
cramped, frosty 18-inch wide gutter. At 1300 the OP

became overt, and we moved into observation and anti-sniper
positions, one man at each end of the gutter between the two
roofs and one onto the rooftop overlooking the square.

We spent the afternoon acting as a communications relay
and also constantly reporting the size and direction of the
marchers and the crowd in the square. In the evening more
troops were moved into Coalisland and eventually the crowd
was dispersed by a baton charge. The security forces then
began to retire, covered by us in the OP.

It had been planned that our withdrawal should be by a
'pig' [armoured personnel carrier], which was to dash in,
pick us up, and then dash out. Unfortunately the first two
sent to pick us up got lost. By this time the crowd had
started to re-form for a meeting. The third 'pig' eventually
got as far as the gates of the mill. Meanwhile we had climbed
down from the roof but we failed to get to the gates before
the collection party was driven off; their position was
untenable as they were surrounded by a crowd about 150
strong.

It was impossible for us to get back on the roof as we
would have been completely compromised in so doing. We
concealed ourselves as best we could in the mill yard. The
crowd continued to increase in size and at 2130 hours the
meeting started. The crowd size was approximately 200–300
at this stage and it was evident that it would be impossible
to lift us safely. We were approximately thirty-five metres to
the rear of the crowd which was pressing against the gates of
the mill yard. It was decided that we cut our losses and
eavesdrop on the meeting.

Initially, Private Smith moved up the yard to attempt to
listen in. He managed to get within five metres of the rear
of the crowd, but had difficulty in making out what was
being said as their loudspeaker was distorted, which together
with the Irish accents made the speeches practically unintel-
ligible. Lance-Corporal O'Riley, being an Irishman, moved
up and took over from Private Smith. At this stage the crowd

was in an extremely ugly mood as the subject on which they were being addressed was the 'destruction of the security forces'.

The meeting eventually closed at 2210 and the crowd began to dribble away; Lance-Corporal O'Riley returned to give me a report on the substance of the meeting. Private Smith returned to the position by the gate. At this stage a car drew up by the gate and fired six rounds into the yard, all shots going high, possibly at the OP on the roof where we had been some fifty minutes before. Unfortunately, Private Smith was unable to get a clear shot at them as other cars and people were still in the way.

At 2225 a 'pig' eventually got down the side road by the mill. It was driven on this occasion by a member of the recce platoon and because of his determination to rescue his colleagues we were lifted back to base. So ended our first day on operational service in Northern Ireland.[15]

Despite the fact that this incident in Coalisland had not led to any fatalities, the violence was steadily increasing in Belfast and Londonderry. The Royal Marine sniper from 40 Commando continues his narrative:

My company was posted to the northern part of the unit area, the sniper section being split-up back to individual troops. After a few days in my troop I was moved onto the newly formed company intelligence section. We operated alone with our own transport, and on our first trip into town one evening I was amazed at the amount of shooting that was going on. It reached a climax one day when rumours were rife that there was going to be a short cease-fire, and intelligence reports indicated that the IRA were really 'going to have a go' until the deadline.

I can't remember how long this 'cease-fire' lasted — 24 hours I think — however, intelligence sources indicated that the IRA were going to force the security forces into breaking

it and have another go at us. Myself and (Phil), another
sniper from my company, were sent down to Brown Square
Police Station in the very south of our unit area on our first
sniper task. As we entered the police station in an open Land
Rover driven at high speed, a group of youths stoned us.
However, I was gratified to note one youth threw a stone,
the size of a half house-brick, which hit a lamppost and
bounced straight back and hit him on the head. Our briefing
lasted about twenty seconds, as I recall: 'Go up those stairs,
you'll find a sangar [defensive position] on the roof, see you
in the morning.'

Our protection consisted of about fifty sandbags piled on
a small wall with the sloping slate roof angled up at the
immediate rear of the sangar. I think there was a 'wrigley
tin' [corrugated iron] roof, and it was very cramped. There
was only room enough for one at the front 'business end', the
other had to squat on the roof immediately behind. The
position faced north and was totally exposed to any hostile
fire that should come our way from the hundreds of windows
in the Unity Flats on the right and Denmark Street Flats
(under construction) on the left. We found a field telephone
but when we tried to test it we were told to 'Shut up,
unless you've got something important to say!' As we were
setting up we could hear small-arms fire echoing over the
city, and being so high up it seemed every shot was being
fired at us.

For several hours we took turns either in the front of the
sangar, sometimes leaning right out to get a better view of
what was happening down the Shankill Road, or hunched at
the rear with only a small field of view. The small-arms fire
rarely stopped for more that a few seconds and we quickly
learned to recognize the sound of different calibres compared
to the Army SLRs. The streets were relatively empty but it
was obvious that a lot of people had difficulty getting home
from work, and at one stage a group of women became

hysterical when they were caught in crossfire in the streets below us.

We had only brought our L42s [sniper rifles] with us and one IWS [night sight]; we also had no binos, so we used the 32 scope on the rifles for observation. As it got dark we found out one of the major problems with the IWS. After looking through it, one is blind in that eye for a few seconds afterwards. You were supposed to observe through the non-shooting eye but that was often impractical.

Being in the position we were, we expected to see a gunman at any moment, but when we did it took us completely by surprise. It was about 1 o'clock in the morning and the fire had slackened to the occasional single shot, when we heard the sound of several pistol shots followed by the unmistakable 'boom, boom, boom' of an SLR. This was directly to our front but behind the old church at Carlisle Circus which was 430 yards away. I was at the front of the sangar with the IWS pressed to my eye looking for any sign of movement when the same pattern of shots occurred again, only closer.

As the SLR was fired, the figure of a man came sprinting round the corner from Carlisle Circus and stood pressed to the wall looking around the way he had come, with a pistol in his hand held up in classic James Bond style and in full view of me. I almost threw the IWS down and put the L42 to my shoulder. It was not a good position to fire from, and as I had been squinting through the IWS, when I put the scope to my eye I couldn't see.

I stayed on aim and gradually I got my vision back; this only took a few seconds but it seemed like an eternity. The man had changed position to what I think was a doorway, and I put the pointer of the sight on his chest and squeezed off a shot. I quickly reloaded and came back on aim but he had simply disappeared. We both thought he might have gone to ground, so we emptied our magazines into all the likely shadows and the ground around where we had seen him.

We saw no more movement and were relieved for the night. We only gave a brief report to the officer in the Ops Room. I was convinced I had missed the guy, as he could easily have ran at the first shot and disappeared behind Unity Flats. However, towards the end of the tour the unit 2i/c interviewed me. He had a list of, I believe, fifty-six names of terrorists shot by the unit, and after a short conversation put my name next to that of a man who had been found shot dead with a wound through the chest. Several months later, in a conversation with a colleague from the adjoining company, he told me about chasing a gunman who had fired at his patrol, but the chase was rather cut short by a barrage of shots from Brown Square![16]

Shortly after this incident the Royal Marine recalled that the unit snipers were brought together to operate from a central location, which made them more effective. In all, about half the casualties inflicted by 40 Commando on the IRA came from the sniper section.

One of the more outstanding feats of marksmanship in Northern Ireland occurred in Londonderry in 1972, and was a timely demonstration of the usefulness of snipers. A prize-winning Bisley shot, this sniper officer suddenly found his skills in demand:

A Ferret armoured car of the Blues and Royals was ambushed as it approached a roundabout on the Foyle Road coming into Londonderry, and was blown up by a mine. It was a prepared IRA ambush, with an ambush party ready to fire on the Ferret.

I was in the Masonic car park camp, heard gunfire, and took up position in OP Kilo. When I arrived at Kilo I found out what was happening, and we started to look in the area of the gunfire for targets. That's when we saw a car drive down Coach Road. It suddenly stopped in the middle of the road, and two men got out and began unloading weapons. More men arrived to join the ambush and they began to take the weapons from the armoury car.

Since they were armed and were joining the ambush on the Ferret scout car we agreed that they were a legitimate target. Working out the range to them was not a problem as we used large-scale maps; we obviously knew where we were and we knew the ranges of all key reference points in the area. The only problem was that the targets were 1100 metres away, and the wind at that range was a real problem because of the long time of flight. At distances like these, wind and other variables are your real enemy.

Whether it was because of Bobby Sands [an IRA leader who had recently died on a hunger strike], or whatever, there were black flags all over the place. Quite frankly it was like being on Century range at Bisley; I had the flags all the way down, and I could read them all off. So I made my estimation as to what the wind correction was for 1100 metres; although, of course, I was off the scale for the wind allowances in my sniper pocket book, but we had trained ourselves in 1000 metres shooting while on the sniper course.

At this distance you have a time of flight of between two and three seconds, and between squeezing the trigger and the bullet hitting the target, a lot could happen. I'll be frank with you, the first chap I knocked down I had actually got the wind slightly wrong. But because of the time of flight he moved away from the car, and just walked straight into the bullet. And that was it, down and out.

The terrorists just couldn't work out where our fire was coming from, so after being joined by my sniper sergeant we were able to continue shooting at them for some time. Our fire was masked by fire being brought down by two platoons from B Company who had been sent in to rescue the Ferret crew. So at first they assumed they were being hit from further down the road, and kept on coming down. We had one terrorist in the bushes on the side, and we had another wounded under a car. During the engagement I fired some 43 rounds and my sniper partner 40 rounds. Subsequently, from intelligence sources, it would appear that we had knocked down ten terrorists. Seven were credited to me and three to my sergeant.

They reckon my furthest target was over 1300 metres, 1344 meters to be precise. This was a shot I took much further up the road. We were very much at the limit of our equipment, especially in terms of the sight which was only ×4 magnification. Initially no one could believe we could hit them at ranges over 1000 metres, so I was being asked on the brigade net what type of weapon I was using, at the very moment I was trying to take out targets. The CO eventually got them off my back by explaining that we had just won the sniper championship at Bisley and that our skills and weapons were equal to the task.

We continued firing at the terrorists whenever a target appeared. One of them dashed across the road, and we saw him go through a doorway. My sergeant fired a shot but we thought we were too late. Subsequently we discovered that having gone through the doorway, he slammed the door and put his back against it, no doubt thinking, 'Thank God for that!' At that moment the bullet hit the door. When we looked later we saw a perfect oblong broadside-on hole going through; the bullet had taken most of the chap's chest away.

One other problem we encountered was because it was July and we were firing from a sandbagged sangar, every time I fired I kicked up a lot of dust, and I personally very rarely saw through my ×4 telescope what was going on after shot release. But I was able to rely on one of my trained spotters who was calling everything with the ×20 scout regiment telescope and another one who had the Mark V binoculars.

After we had been going some time the IRA did eventually work out where they were being shot from, and decided to do something about it. After all, a sniper would normally fire only one shot and then bugger off to a new position, but I was in no position to do that, anyway I was in a well-protected OP. Eventually they mounted an attack on the OP and fired some fifty shots at us, automatic fire in long uncontrolled bursts. God know's where it went, all over the place. But no one worries about that; it's the very specific shooting that deters you, and that came a little bit later. Then they really homed in on us.

After the attack I moved with my sniper-sergeant to an alternate OP that we had sighted previously; a firing platform on the wall. It had excellent fields of view and fire but provided no protection whatsoever. After firing from there, the IRA then sent someone to take us out from the flank. This was an instance where it is very important that your OPs are mutually supporting. I had some other snipers who, up to now, were feeling very out of it but who were our mutual support and covering that flank. They spotted this terrorist sniper, although, regrettably, he got three shots off at us before they returned fire. One shot hit the wall below us; one went between our heads – I'm quite convinced that I got very slight deafness as a result of that bloody round! – and the other went over the top. The terrorist then started to move round this balcony, whereupon my snipers zapped him and he went over the balcony in good old Western style. He was eventually taken away by his friends wrapped in an old carpet in the back of a van, very definitely dead.

This was a very significant engagement from the battalion's point of view, because we hit them and we hit them extremely hard. The level of IRA activity just dropped. There was this sudden contrast with them running mayhem with the previous unit, and then the first time they try a 'biggy' with us they really get zapped.[17]

At the end of the officer's tour in the Province, his snipers were engaged in a further incident:

Just prior to Operation Motorman – or Car Can as it was called in Londonderry – we were relieved in the line by the Scots Guards. This was in order for us to go back and re-form so that we would become one of the assault groups to be put into the area for Car Can.

At this time there was a practice among the IRA that every time a British Army unit changed round they'd have a go at it. Up to that time we had successfully dominated them, so we left our snipers in position and then paired up each of our experienced sniper crews with the incoming Scots Guards. We knew all the possible fire positions and we waited for them to have a go, and sure enough it happened.

Due to a shortage of room in a protected OP my own sniper sergeant was up on the wall doing some surveillance, while the other members of the team – one of my snipers and a Scots Guard – remained in the OP. Suddenly there was a shot, and the sergeant got hit. In the dark all he could see was that his sleeve was torn and that there was quite a lot of blood. He clamped a dressing on, and handed his rifle to the sniper private, telling him that he would get back to camp to have the wound seen to.

I met the wounded sniper sergeant when he got back to camp, and found he had been shot through the biceps. Meanwhile the sniper private back at the OP – who was only eighteen years old at the time – had been told to gather things up and return to camp, but he took it into his own head that the terrorist would probably come up for a second go. Of his own accord he shifted his position by about 20 metres, and waited. Low and behold, the terrorist came up for a second helping and my sniper smacked him at a range of 200 metres straight in the chest.[18]

Of course, not all British sniping was so precise or lethal. During a tour in 1976, 2 Para came up against the North Louth Active Service Unit of the IRA. The paratroopers conducted patrols in dangerous areas in order to draw fire deliberately from IRA terrorists. Three sniper teams from 2 Para were covertly deployed to respond to any IRA attack. A corporal describes how the IRA ambushed one of these decoy patrols:

> Wham, wham, about eight or nine rounds come at us. We're all scurrying all over the place, but we still haven't picked up where the shooting is coming from. We contact the sniper party, and one of them reckoned he'd seen something, so they put down a couple of rounds and said they'd got a kill. The same night we went up to this great big hill of gorse and found two dead cows the snipers had shot![19]

Despite such incidents, British snipers played a major role in defeating the IRA's 'shooting war' strategy. By the end of the 1970s the IRA had largely abandoned the use of firearms against the British Army, and relied primarily on the bomb – an indiscriminate weapon, but one easier and safer for them to use. Increasingly, the IRA came to fear to arrival of snipers in their area, as this comment from a sniper NCO suggests:

> The sniper spreads fear in the enemy. A graphic example of this happened on a patrol through a particular city centre in Northern Ireland. We went past a well-known terrorist haunt, and there were three or four suspected terrorists lurking around outside this building – as we expected there would be. They didn't bat an eyelid at any of the patrol, until I came around the corner with an L96 [sniper rifle].
>
> There was visible panic: three of them went indoors while one stood outside and watched. He then sent some local kids over – a well-known trick of theirs – and they asked lots of questions about me. I just didn't answer or gave them loaded answers, to further the fear. We were followed for the entire patrol. That one sniper rifle caused an amazing stir. The terrorists were not happy that it was there. It was something they hadn't seen before; it was new and it was a threat to them.

There were occasions, later, when we were on border patrols, where we allowed suspected terrorists to see the men carrying a sniper rifle. The word would then get around that there were qualified, professional snipers out and about, who were prepared to do their stuff. I will never know to what extent that fear carried through, but I like to think it played its part in making our tour a very quiet one as far as terrorist activity was concerned.[20]

The border areas of Northern Ireland provided some of the most fruitful areas for IRA activity. Not only was much of the local population staunchly Republican, the nearby Irish Republic allowed terrorists to escape the British security forces. The British sniper NCO describes his deployment along the border:

One of our main tasks was border patrolling and the closure of border crossing points once they had been opened by either terrorist organizations or local farmers. In the past, patrols had been shot at by terrorists, therefore all angles had to be covered. I managed to convince my company commander that snipers would give him a long-range precision weapon that would cause little if any collateral damage or injury to innocent bystanders — as some of the heavy belt-fed weapons we use would do — and cause a backlash from the civilian population. And he also had a trained observer.

Luckily, the company commander was sold on this idea. Depending on the type of operation, I ended up with between four and six sniper pairs under my command, to deploy as I saw fit. Not only did it give me a chance to get sniping in the forefront, but it gave me a chance to learn and to deploy snipers operationally.

The company commander gave me a helicopter on several occasions to go and do some air recces of the positions we were going to. And that consisted of a fly past at some distance, so as not to alert anyone watching the helicopter as to the area of our interest. In conjunction with maps and air photographs, I could look at the terrain and decide the likely positions where I could deploy my snipers. Then I would sit down and work out my plan alongside the company commander.

A blue-on-blue or friendly forces contact was obviously to be avoided. So in the orders process, I would stand up and brief the commanders on the deployment of the snipers: what would happen and what they would look like. I used to get one of the snipers in on the orders group, in full ghillie suit with his equipment, to show the commanders there, especially the younger ones, what a sniper would look like, so there was nothing that could lead to somebody shooting at one of my guys.

On one particular occasion, there was a small bridge that had to be blocked, and the only place for us to give covering fire was very close to the bridge. So one other sniper and myself went in early, stalked forward to a position where we could overlook the bridge and awaited the arrival of the engineers. But before they could start work the bomb disposal guys had to come in and clear the area to make sure it wasn't booby-trapped.

It was actually very satisfying to watch them doing their job, listening to them chatting only a couple of feet away from us. And it wasn't until one of them came very close that I quietly said to him, 'Just let the rest of your guys know we're here. We don't want anyone over-reacting and giving us away.' And the look of shock on his face was a picture. The guy had no idea we were there. It was a good boost to our pride.[21]

During the early 1990s the IRA conducted a cross-border sniping campaign, killing several men in the security forces. When it was discovered that some of the shots had been made by .5in calibre Barrett heavy sniping rifle – capable of piercing light armour and with a maximum range in excess of a mile – there was great public interest. Rumours spread that the sniper was a renegade

from the US Special Forces, but this was little more than wild
speculation. One revealing comment about this affair was made
by a British sniper authority: 'What is interesting about the .5in
IRA sniper is that the ranges were all within the effective
capability of a much smaller calibre rifle, no shot being over 400
metres. The rifle was used for psychological and propaganda
value.'[22]

Despite this, the IRA threat was taken seriously by the British
Army and precautionary measures were adopted, as the sniper
NCO explains:

I was approached by another company commander to do a
counter-sniper study. He had several permanent border
crossing points which his own men were occupying. There
had been a history of shoots against these posts from certain
positions, and he'd heard of the use of our snipers on the
border-closure operation.

He put a vehicle at my disposal, so that a corporal (an
experienced sniper) and myself went out in civilian clothes
in a plain car along the border areas between the different
bases, and carried out a full counter-snipe study. We had
been given the firing points, dates, times of day that the
shots had been taken against the security forces. A pattern
had been set, and it was a starting point from which to work.
So we then went out with cameras – and with sidearms – to
these areas and had a look at them in daylight.

Positions were found for all the bases that we were asked
to go and look at. These positions were then visited at night
under cover of darkness and in the early hours of the morning
– in uniform – as part of a normal 'green army' patrol. We
then ran a few small operations where a couple of the
positions were 'proved', i.e. we went in one night and
occupied and confirmed we could see the arcs of fire we
required. I then did all the paperwork, the marking of maps,
the air photographs, the write-ups, the reports, etc., on these
areas – and they were kept on file.

There was also an operation put into place where if the threat was to go up, we could be called for and deployed at very short notice to occupy the positions for a possible counter-snipe. We were, in fact, deployed on a couple of occasions when the threat went up and a shoot was expected. But, unfortunately, nobody came in to take a shot against the security forces, so therefore we never actually got to follow the plan through. With the exception of actually pulling the trigger, the operations worked very well. We were never compromised, and these positions are obviously still viable should somebody wish to use them. Hopefully, that option will never have to be used.[23]

Discussing the objectives and tactics adopted by snipers in the 1970s, this officer sums up how useful sniping could be within the infantry battalion:

By carefully siting our OPs, we were able to achieve a form of domination. We were able to put in the mind of the terrorist the certainty that if he tried something he would get shot.

Our whole system was very well integrated. We each had two radios, one on the battalion net and one on the company net. Not only were we able to provide the battalion commander with information, we were able to support company commanders from our various OPs because we knew when their patrols were coming out. We would then provide cover for that particular patrol. Generally patrols knew that they had sniper cover while they were on the ground. Unfortunately, this level of integration was probably unique in Northern Ireland. We had spent a lot of time training our snipers, and we thought we were using them to the best advantage. When they weren't shooting anybody they were acting as eyes and ears but always having the capability to provide armed cover.

As when using any other specialist, you had to have a

very good understanding of the skills of snipers, in order to use them to their best advantage. I was fortunate as Recce Platoon commander in being a trained sniper, and I was very lucky that the CO gave me a total free range. The overall intention was to dominate with my snipers but how I actually got on with it was totally left to me.[24]

Chapter Eight

Training the Sniper

To become a sniper in the US or British armed forces, the recruit has to pass a demanding course lasting from six to eight weeks. Only experienced infantryman with good shooting skills are accepted on the course, and even then the drop-out rate is high.

The US Marine Scout Sniper School at Quantico in Virginia believes in hard training, as Captain Tim Hunter, a former officer in charge, explains:

Our students are on the go all day long in a combination of field and classroom work. But they still have a lot of studying to do at night because they are tested the next day. And they also have written assignments. This is not a boot camp and we don't treat our students like recruits. They're already good Marines or they wouldn't be here. But we make the course stressful. Being a scout sniper is an extremely stressful business. The school shows us how a future scout sniper will act and react under pressure. If he can't handle that stress in a school situation, there is no way he will be able to cope with it on the battlefield.

We don't tell our students what score they must have to graduate in the field skills portion of the course; we want them to give it their maximum effort. We don't tell them how well they're doing because we don't want them to know that they may have already passed this part of the course in the fifth or sixth week and then take their packs off. We

want them to do their best to the very last day. If a Marine
thinks he can come here and 'hot dog' his way through the
school, he'll be out on his ear.[1]

Apart from the obvious emphasis on shooting skills, great
attention is paid to field craft. Each student must conduct a
number of stalks, crossing over eight hundred yards of terrain
under constant observation, then fire two blank rounds at a hidden
target without being detected. Gunnery Sergeant Paul S. Herrmann, a former NCOIC (Non-Commissioned Officer in Charge)
of the school, describes the process:

> We really stack the deck against the students. The most
> dangerous thing a sniper can face is another sniper, because a
> sniper knows just what to look for. Our students are facing
> two trained snipers [both instructors at the school] who act
> as observers during the stalk. The students are heading
> toward the observers, down a channelled area, from a known
> distance. They have a four-hour time limit in broad daylight.
>
> Coming to within 200 yards of the observers, who have
> high-powered spotting scopes, the students must set up, fire
> a blank round at them, and then wait. If the observers can't
> see where the first shot came from, then the student fires
> again. If his position isn't spotted by then, the student has
> made a 'possible' on that stalk. That's very rare, especially in
> the first weeks. The other instructors and I pride ourselves
> on being good observers.
>
> If you don't think there's a lot of stress involved, try
> crawling 800 yards through the brush, firing a shot, and
> before you get to fire the second round, being spotted by the
> observer because of some silly mistake. In a real situation, of
> course, the sniper would never come that close to a target.
> We believe if they can set up on us, with the deck really
> stacked up against them, they can set up on anyone.
>
> The [M40A1] rifle we use is hand assembled at Quantico.
> It is based around a Remington Model 700 receiver and it

fires the 7.62mm NATO round (.308 calibre). A few former
Marines tried to build one but they figured it would cost
about $3,500 to put it together, although it doesn't cost the
Marine Corps that much.

It's not a very pretty weapon, but you can drag it through
the bush and give it one hell of a beating and still hit a
target at a thousand yards. We work the students to achieve
a 'one shot, one kill' ratio. They fire from the 300- to the
800-yard line at both still and moving targets. At the 1000-
yard line, the targets are stationary.[2]

Since retiring from the US Marines, Carlos Hathcock has frequently lectured to sniper recruits. On one instructional course
with US Police snipers he made these practical points, relevant to
any type of sniper:

When you move into an area, move slowly – slow, deliberate
movement, observing everything you can see. Analyse the
terrain and know how you're going to get there and how
you're going to get back. Never get into a hurry if you don't
have to. After you occupy your firing position, you have to
sweep the area in front visually, making sure there is nothing
that will endanger or affect you or your partner. Scan the
area from left to right, then right to left. Note the bushes,
trash cans, trees, cars and so on.

You and your observer set up a range card so that you
know exactly what is out there by writing down each prime
point. You number each obstacle and point so that you each
know what to shift to if a number is called out. When the
observer says, 'Section A, position 2,' you know exactly where
he is talking about and you don't have to search the whole
area for the bad guy. If you are watching a building, number
the windows and you'll know exactly where the observer
spots the suspect when he calls out 'window three'.

Keep a data book with the information that concerns
your rifle. Know how many shots have been fired through

that barrel. If you wear out a barrel on the range, it would
be nice to know that you need a new barrel before you have
to make that one well-aimed shot for real.

Log books contain the basic information and record of
each situation you have been involved in. Keep a record of
Who, What, Where, When, Conditions, Wind, Range,
Shots Fired, Situation and Location. The information is
invaluable later to your intelligence people or command post.
You are in the best position to see what is going on.

If you are going to train to shoot people, then targets
should look like people. Few suspects dress up to look like
bull's eyes. When you look through that scope, the first
things that jump out on you are the eyes. You see a living
human being in your sights. I teach with targets that show
life-size faces.[3]

The requirement to make targets more realistic for sniper-training
purposes has been common practice among many instructors.
Chuck Kramer, an American training Israeli snipers, outlined his
methods:

I made the targets as human as possible. I changed the
standard firing targets to full-size, anatomically correct
figures because no Syrian runs around with a big white
square on his chest with numbers on it. I put clothes on
these targets and polyurethane heads. I cut up a cabbage and
put catsup onto it and put it back together. I said, 'When
you look through that scope, I want you to see a head
blowing up.'

I had about fifty applicants for the course. We took
the youngest guys who had most time to serve. I asked all
the right questions. 'We want you to look through and
just see a target and watch his head being blown apart.
Are you going to do this?' If there was any slight hesi-
tation, I'd tell them to go home. I didn't have time to fool

around. We shaved it down to fifteen guys because we only had fifteen rifles at that time.

We built our own 1000-metre range on a desert plateau. Anything I wanted, I did. I drew 2.5 million rounds of ammunition and we started shooting. I told the students, 'If you guys won't kill anybody I aim you at [with the] first round, then out.' There was an esprit de corps because they were getting good at this.[4]

The psychological requirements necessary for a good sniper were explained in comments from a British sniper trainer in the Parachute Regiment:

One thing we had to consider when we picked snipers was how they would react to shooting a target they could quite clearly see was a human being, and not just a grey blur that is already threatening the firer. We looked for a degree of 'compartmentalization' in our snipers. He might see someone shaving, standing in a meal queue, or whatever, but he had to see them as the enemy. As far as we could, in training we used targets that were not obviously military or threatening.

The average paratrooper tends to be a bit more aggressive or more rugged than normal. As an example of their black sense of humour, we went down to Hythe on the OP range [constructed to recreate urban conditions in Northern Ireland]. There are moving targets on each street, and you start with very simple black and red targets of which the red has to be engaged. The idea is that it is progressive, so that you can eventually distinguish between targets and identify the one that appears carrying a gun. We had one pair of snipers in the roof overlooking a target of a woman and a pram. The woman was shot, and the company commander running the range said: 'Well, that's another orphan child to add to the problems of Northern Ireland.' At which point there was another shot, and it went straight through the pram.

This is not to say that some paratroopers didn't suffer their own after-effects. I talked to them and on the whole they were honest, and I think they realized it wasn't for them. There wasn't so much status in the job that it was something they were going to lie about, so they'd say, 'I don't think I can do that.' And that was fine. There was no discredit to you, it was just that we were looking for someone with a different outlook.[5]

Another British soldier, who served in Northern Ireland, discussed some of the moral aspects of sniping, and the type of man who made a good sniper:

Ultimately, for general training there is very little difference in training a chap for a reconnaissance role and training him as a sniper. All the basic skills you need are the same, other than shooting. For both aspects you're looking for a guy of above-average intelligence, and the only way you can justify having that sort of chap in that sort of role is in a very specialized role like the recce platoon.

Sniping is all about patience, something akin to fishing. One of my best snipers was one of my youngest ones; he could sit there for hours on end – rather like the guy sitting on the canal bank with his fishing rod, with his finger on the line, who is relaxed in one way but totally alert in another. The minute he feels the line tug, he's on.

Inevitably there is a sort of type of man who is attracted to that sort of role; the older private soldier, for instance, who hasn't settled desperately well within the constraints of normal company discipline, but thrives well in the recce/sniper environment. It's not so much the oddball that makes the good sniper, but I'm quite convinced that there is one prime characteristic that is necessary, and that is the hunting instinct.[6]

Without exception, sniper experts are agreed on the importance of the hunting instinct. Some recruits have long experience as hunters, but others need to have their hunting instinct developed further. This can be done with realistic training. One such training exercise was observed by journalist John Coleman at the sniper school of the XVIII Airborne Corps. The five-week course is attended by around thirty candidates, who undergo a rigorous selection procedure. In the words of the sniper trainers, the aim of the course is, 'to provide specially selected personnel, who are highly trained in fieldcraft and marksmanship'. They must be able to 'deliver long-range precision fire on selected targets from concealed positions in support of combat operations'. Coleman describes the course of the exercise in detail:

By 1100 hours the hot, sodden air had settled oppressively over the thickly wooded forest; the sun already burning its way through the green foliage to bake the forest floor. Call sign 'Alpha Charlie', the two-man sniper team from the 82nd Airborne Division, felt the new day's soggy heat flow into their underground 'hide' through the loopholes facing their target area. The stale air in their bunker didn't stir, it just became more stagnant, more uncomfortable, with each passing minute.

The 6 × 6 × 4ft hide had been their home for better than two days, sharing the cramped space with their combat-loaded rucksacks, load-bearing equipment, radio, camou-flaged 'ghillie' suits, sniper weapon system and M16 – and a variety of insects and crawlers. Sleep had come only in snatches, one man observing the target area – a small bridge some 300 metres away – while the other rested.

They were already tired before they'd even started their target area observations. The move-in had been a slow, patient and painstaking process: they were behind enemy lines, and compromise could come from the slightest mistake. Then there'd been the all night digging and camouflaging session of their hide. It had to be letter perfect:

deep enough to stand comfortably, large enough to provide at least minimal comfort for a number of days, situated to provide optimal observation and weapons' coverage of the target area, strong enough to provide overhead cover and, perhaps most important, so well blended into the terrain that even the most experienced enemy scout would pass it by. All done at night, deep inside enemy territory.

PFC Paul Normandeau leaned against the earthen ledge of the hide, an eye glued to the M49 spotting scope. Periodically he'd change to the binoculars, sweeping the area while giving his eyes a rest. An occasional wave of sodden air wafted through the loopholes, cooling for a brief instant the sweat forming on his black and green camouflage-painted face. Never for more than a few seconds did he take his eyes off the target area. Their mission was too important to allow even a moment's slack.

He swept the bridge and the road leading up to it for what was probably the hundredth time. Movement. On the road, maybe seventy-five metres on the other side of the bridge. He looked again. Two figures, carrying weapons. One dressed in the uniform of a Soviet officer, the other in a green and sand-spotted camouflaged jumpsuit. Normandeau nudged his teammate, PFC John Labis, with his foot.

'What's up?' Labis asked quietly, rolling on his makeshift bed of ghillie suits to face his teammate.

'We've got the primary target coming down the road. He's got a buddy with him.'

Labis stood up and peered out of the second loophole. He spotted the two distant figures immediately, then peered again through the ART-2 scope mounted on his sniper rifle. The two men jumped into clear focus. 'Soviet uniform' – an adviser to guerrilla forces operating in the area – was definitely their primary target. He was carrying intelligence information and documents which could affect future US operations in the area. They hadn't been briefed on 'spotted suit', but he was probably a senior member of the guerrilla forces. A good secondary target.

Labis kept the Soviet centred in his sight as the two moved closer to the bridge. It was a known range: Labis adjusted his scope for 300 metres. Normandeau fed him wind conditions – nil. It should be a good shot. The two enemy troops closed on to the bridge, half alert while scanning the bush alongside the road, but generally unconcerned. They were in friendly, guerrilla-controlled territory.

The Soviet officer crossed and then stepped off the bridge. One shot cracked through the still, hot morning air; the officer crumpled onto the dusty road. Acting out of instinct, the guerrilla dove for cover. Then he remembered the documents on the adviser. They had to be delivered. He crouched, then dashed over to the inert body on the road. Frantically he pulled at the pouch on the officer's belt. A scant moment later a second shot rang out, and the guerrilla dropped sprawling to join his comrade under the washed-out, pale blue sky.

Callsign Alpha Charlie had carried out its mission.

'OK, sniper. Come on out of there.' Sergeant Scott Raitt, NCOIC of XVIII Airborne Corps' ATMU (Advanced Marksmanship Training Unit) Sniper Committee, delivered his order to the empty woods.

To a casual observer, Sgt Raitt was talking to a piece of Fort Bragg's pine needle-strewn forest floor under his feet. Then, a 2 × 2ft section of the forest floor broke away, lifted up and was pushed aside. PFCs Normandeau and Labis climbed out of their hide to face a tough critique of their mission and hide construction.

'You accomplished the primary mission. Good,' Sgt Raitt told the two sniper students while marking down points on his clipboard. 'Let's look at your hide.'

For all appearances, the students' hide was undetectable. The small mound forming the roof was unnoticeable, and their camouflage blended in perfectly with the terrain. Yet there were errors which an experienced guerrilla – especially one familiar with the area – or a sharp enemy special unit operative would catch. A bit of spoil, the sandy clay-like earth dug from the hole, had been left uncovered by the road. The position of a few tree branches at the rear of the hide looked man-made rather than the result of natural fall. Grooves in the ground in front of the loopholes were a shade too pronounced, and the loopholes themselves had collapsed slightly from the weight of the overhead cover, thus restricting vision. Small points that could lead to compromise – or worse – in combat. Here, they led to the deduction of critical, graduation-required points.

Sergeant Raitt and his staff finished their inspection. Raitt explained the errors to the students, using them as teaching points. Then came the verdict: 'Good hide, snipers. Well done.'[7]

The overriding factor in the success of an operational sniper mission is to put together all the different things a sniper has learned during basic training. A British sniper NCO provides this account of a mission, from the perspective of the man who is literally on the ground:

When carrying out a sniper mission, first there is an intense planning period, where you sit down and look at maps and air photographs, intelligence updates, etc. Once a plan has been worked out, you then get to the deployment stage, and – depending on the type of operation – you are inserted by foot, vehicle or helicopter. Once you've deployed, initially you feel an overwhelming sense of being alone – it's just you and one other person. You feel very small and vulnerable. But after a few minutes, once you settle into the patrol and realize that you are better trained than the enemy, then it starts to become challenging and enjoyable. It is that tense excitement that something could go wrong at any moment that keeps you on your toes.

With every footstep you make, you try not to leave ground sign, try not to make a noise. There should be no requirement to speak, although you can make the odd whisper into each other's ears. Generally, you keep the noise levels down.

If you get to the crawling stage, where you've got to cross a piece of real estate on your belt-buckle [stomach], the trick is to pick an object that you know you will be able to see from both standing and belt-buckle levels. Using it as a sort of aiming mark you can maintain the right direction, because when you're down at belt-buckle level, the world seems a very small place. At best, you can't see further than a couple of feet in front of you.

Except on the odd occasion, you don't notice pain. Bramble bushes rip you to pieces, they go straight through your combats; rocks will graze you, and you'll get poked in the eye by grass and twigs.

While stalking, you must stop to listen. Your ears are a life saver when you can't see very far. You must notice what is going on around you in nature. Birds will make a different type of noise when there are humans rather than animals in the area. Every animal has got its own alarm sequence. The more experience you get, the more you notice these things. All these thoughts go through your head while you're stalking. You're trying to find the one thing that's going to give you an advantage over your enemy. All the time you're trying to remain quiet, keeping your senses on maximum alert.

Once you've stalked to the position and found your target, you then spend time preparing that position. You camouflage it and camouflage yourself to look like the position you are going to fire from. You may have already changed the natural cam [camouflage] on your ghillie suit three or four times on the way in – depending on the changes in the vegetation. A good ghillie suit is covered in elastic, and in that elastic needs to go all the natural cam for the area that you are crossing.

Once you have done all your cam and got your weapon into position, it's up to you and your partner to take the shot. You work out the range, windage, if there is any, and confirm that you are both happy with the target. Then you decide the time to fire. Once you've taken the shot, it's like sending a postcard to say you're there. Everybody now knows you're in the area, and they are going to do their damnedest to look for you.

The extraction has to be better than the infiltration phase. On infiltration, the enemy didn't know you were coming, on the exfiltration they know you're there. The enemy may react in several different ways. They may all go to ground through fear, they may fire wildly in your general direction, or they may follow up and try and catch and kill you. If the latter, your exfiltration must be good. Rule of thumb says you don't go out the same way as you came in, especially in case you've been compromised by an enemy patrol that's been moving behind you, found some ground sign and is now waiting for your return. The extraction must be taken with the utmost care and attention to detail.[8]

Once a soldier has graduated from sniper school, he continues to learn and improve his knowledge of sniping. As one sniper put it: 'No one man has a monopoly of good ideas. The learning process never stops.'

When the sniper returns to his unit he must put his new-found knowledge to good effect. This often means convincing company and battalion commanders of the usefulness of snipers to the unit. Knowing how best to deploy and organize snipers within the battalion remains a problem. A typical eight-man sniper section is probably too small to exist in its own right, but snipers disagree over who should act as the parent unit. In the US Marine

Corps, for example, the section is located within the surveillance and target acquisition (STA) platoon of the headquarters and service company, but some sniper experts believe that the STA's concentration on reconnaissance undermines the sniper's shooting role, making him too much of a scout and not enough a sniper.

Until the mid 1980s, snipers in the British Army faced more serious organizational problems. A paratroop officer who trained his unit's snipers in the period leading up to the Falklands conflict had to circumvent some of the traditional barriers that prevented the correct deployment and training of battalion snipers:

> I was very much of the opinion that snipers should be lodged with the intelligence cell but should work directly to the battalion CO, and not to a company commander or the intelligence officer. The problem was that there was no establishment for a sniping platoon within the battalion. But snipers had to 'belong' to somebody, and at one point our snipers were drummers – that's not uncommon – and what then happened was that somebody who could play a drum became a sniper – which isn't necessarily the best qualification to become a sniper! And we had some snipers, men who were good shots, who had to learn to play the drums, which again was lunatic.
>
> We got round that one way or another, and we had an effective working system whereby our snipers – who were the nucleus of the battalion shooting team – trained as a platoon. In normal circumstances they worked with the training wing, providing a base of smallarms training, so that, for example, when companies wanted to run a small arms instructional cadre, the snipers provided the bulk of the instructors. They also ran remedial cadres for soldiers who were having difficulty in being effective as riflemen, and they even ran basic sniper cadres, so, at least, they could bring on fundamental infantry skills, like concealment, movement, signalling. They were all mortar fire controllers, and many of

them I managed to get trained as artillery fire controllers, as
I felt it important that, if necessary, a sniper could bring
down artillery fire onto a target.[9]

One other continuing argument is the choice of weapon. Should a
sniper be armed with a bolt-action rifle – such as the US Marines'
M40A1 or the British L96 – or use a semi-automatic rifle, such as
the US Army's M21 or the German PSG-1. At present the debate
has swung in favour of bolt-action rifles, with the US Army
turning to the bolt-action M24 system, but a substantial number
of sniper authorities maintain that a semi-automatic weapon
would be better for general military use.

A related problem is the argument over the type of weapon to
be carried by the observer. Should it be a 7.62mm rifle of the
same calibre used by the sniper, or should it be a 5.56mm rifle
with high fire-power but a short range. Within the US Marines,
the observer's weapon has alternated between the 7.62mm M14
and the 5.56mm M16. Sergeant Brian Poor, a Marine sniper and
sniper instructor, has argued vehemently for the employment of
the M16, which is the current armament of the observer:

> The mission of the observer is twofold. First, he locates and
> indexes the targets to the sniper, reads the wind and observes
> the strike of the round. Very few 800-plus-yard shots are
> first-round hits. The observer attempts to note the strike of
> the round, give an adjustment, and get the sniper on target
> for the second round.
>
> Second, and most important, the observer is responsible
> for protecting the sniper whose concentration is on the
> target. All of the training, tactical knowledge, and camou-
> flage cannot guarantee that a sniper team will never be
> compromised. When a well-camouflaged sniper team is
> compromised, it is not at 600 yards; it is usually a chance
> contact, often at night, almost certainly eyeball-to-eyeball
> distance by a much larger force. Precision rifle fire from a
> second scoped rifle is at that point of very limited value.

What is needed is a high volume of *accurate* automatic fire to produce enemy casualties, temporarily stun the enemy, and buy the sniper team a few seconds to break contact.

An example of the value of instantaneous, high-volume fire response took place during the third night of the ground war in Kuwait [during Operation Desert Storm]. STA Platoon 3/9 was travelling in three HMMWVs [light vehicles] though a fruit orchard near Kuwait International Airport. The orchard was heavily fortified with bunkers, dug-in armoured vehicles and concertina wire, but appeared to be abandoned. Our orders were to push through and seize the airport. We were engaged by an [Iraqi] RPK machine-gun firing from a bunker twenty yards away. The machine-gunner only got off about fifteen rounds before he was killed by a high volume of automatic M16 fire directed at the muzzle flashes by myself and other observers. Would scoped M14s have worked? Certainly not as quickly. The gunner was dead three seconds after he pulled his trigger. Had he lived five seconds, I most likely would have been hit, as his last couple of rounds passed inches from my head.

A sniper observer should carry an M16 with M203 grenade launcher attached. If you want two sniper weapons firing, then send two sniper teams, but do not make a sniper team's already very dangerous job worse by sending them out without an *effective* automatic weapon to protect themselves.[10]

Discussion over weapons, organization and deployment will continue, reflecting different sniper preferences among the armed forces. What is now sure, however, is that the sniper's role is much better understood than formerly. Describing this role, Marine Sergeant Gary Grella said: 'Snipers are a self-sufficient, low-cost supporting arm. A two-man sniper team is capable of stopping or delaying a company-sized force.'[11] This view was echoed in a comment by a British sniper, summing up the sniper's position within the wider scheme of things, but with an important proviso that snipers should be aware of the importance of the 'low-cost' factor within the tactical equation:

> There will always be the need for effective rifle marksmanship, and the sniper is the apex of all infantry skills, and used properly he should be an asset in war, and an enormous training asset in peace. But rather like anything else, if you make the sniper a prima donna, you will then begin to erode effectiveness, becoming counter-productive in terms of cost. In some ways it is better that the sniper doesn't get too many freebies or too easy a time, otherwise he gets resented by the others.[12]

Chapter Nine

Sniping in Contemporary Conflicts

In the 1980s, after decades of official neglect, snipers finally achieved regular status within the armed forces of the Western world. Previously, snipers were raised, trained and equipped on an ad hoc basis during hostilities, and then disbanded. In the new system, not only were sniper sections organized on a permanent footing, but battalion and company commanders began to accept and benefit from the special skills a sniper could provide.

Since the early 1980s, Western nations – notably the United States, Great Britain and France – have been involved in the deployment of peace-keeping forces. Peace-keeping duties impose severe restraints on soldiers. Often, they find themselves acting as targets for local warring factions, but the peace-keepers know that vigorous return fire will usually cause collateral damage, further inflaming an already delicate situation. Here, at least, the sniper can lessen this problem by eliminating only armed and threatening individuals. Despite claims made by other branches of the military, it is the sniper who is the most capable of conducting a surgical strike.

This fact was underlined in the US involvement in the Lebanon. In 1983, US Marines – along with French and Italian troops – were despatched to Beirut in the ultimately hopeless task of attempting to administer a cease-fire. The Marine bases in Beirut came under repeated attack from Muslim forces who were determined to expel the Americans. The fighting in Lebanon had been characterized by extreme savagery and not by any tactical finesse. Consequently, the Muslim militias were doubly surprised when Marine snipers were eventually permitted to return fire. As a captain who served two tours in Beirut, R.D. Bean was in a good position to observe the sudden and dramatic arrival of the sniper on the Lebanese battlefield:

> Who could know which faction the enemy represented, and what difference would it have made to know? They fired at the Marines with assorted small arms from a hundred vantage points at sometimes absurd ranges and with annoying regularity. They particularly enjoyed appearing on the flat, concrete rooftops so common in the city.
>
> The sniper team noted a partly uniformed enemy wearing a cloth around his head and carrying an AK-47 appear on a distant rooftop. Holding his weapon aimlessly he emptied a 30-round magazine towards the Marine positions. At 800 yards he probably felt quite safe in exposing himself to the 'Great Satan'.
>
> The bullet sprayer stayed in sight for too long. The sniper steadied for a long instant, and fired. 'Decorated' by a match-grade .308, the enemy crumpled and lay unmoving, still in view, his rifle fallen aside. The snipers waited.
>
> Sure enough, the enemy's partner stepped into view. Shocked by the sight of his downed companion he hesitated whether to look closer. But after all, the range was so distant a moment's exposure could not matter. Surely, he had been felled by a random bullet. The Marine sniper again touched his trigger, and the second enemy collapsed to lie without movement beside the first. Two shots, two down and out.
>
> The sniper relaxed behind his M40A1 [sniper rifle], but not for long. His observer said, 'Oh, oh, there's number three.' And there he was, a third enemy, apparently disbelieving what he saw, turning to look west towards the American positions, raising a clenched fist in silent rage. Crack! The 173-grain projectile took him squarely in the

chest, buckling his knees, bending him into praying before
he fell limply on to his side.

A fourth enemy stepped into view, examined the bodies,
looked about, apparently in wonder, and died, to lie atop the
others forming a monument to USMC scout sniper death
from afar. The Marine snipers waited patiently, but the
bodies and the AK-47s remained undisturbed until dark hid
them from view.[1]

Corporal Tom Rutter arrived with the US Marine 1/8 Battalion
Landing Team in June 1983. Rutter, and his fellow snipers, had
been ordered not to return the incessant fire coming from the
Muslim gunmen in the warren of houses known as Hooterville,
which faced the US Marine base. Their frustration was relieved
when the order was eventually rescinded. Working with sniper-
team members, Corporals Baldree and Crumley, Rutter describes
their first action:

I had just assumed watch on the roof with Corporal Baldree
and was making my first sniper's log entry after examining
through binoculars the parapets and fortifications that surrounded us. '1730 hours,' I wrote. 'All quiet. No new
obstacles or positions.' I peered out. Hooterville was almost
dark, although all the light had not yet drained out of the
sky. Maybe the ragheads had already heard we had turned
and were going to shoot back.

Suddenly, a string of bullets screamed over our heads
above the sandbags as a heavy machine-gun opened up from
somewhere at the edge of Hooterville. I ducked instinctively.
So did Baldree. We grinned sheepishly at each other as the
next burst of machine-gun fire chewed viciously at the
sandbags.

We darted from one firing port to another, taking quick
looks out, trying to locate the nest. Every few seconds the
machine-gun snapped a few rounds at us. The rounds were

hitting near the firing openings in the sandbags, but so far none of the bullets had come through. I didn't want my head framed in an opening when the gunner finally hit it.

'I don't see a goddamned thing,' Baldree cried in exasperation.

'Wait . . .' I said. 'Okay . . . The tall building at two o'clock. See it? Second floor. The dude knows his stuff. He's not firing from the window. He's back in the shadows.'

'I see him!' Baldree hooted as quick, flickering muzzle flashes lit up the room. We were watching through scopes and binoculars.

'He's too far back in the room,' Baldree said. 'We can't get his angle from here.'

I had a thought. 'No, but we can double-team the shitbird.'

I grabbed the field phone and got Crumley who was on watch at the grunts' concrete barrier on the ground floor. I quickly outlined the plan to him. Baldree and I would mark the window with rifle tracers, if the grunts would follow up with their M60 machine-gun. We finally had a chance to fight back. I didn't want our first raghead getting away.

I checked the range with the mil-dot scale in my rifle scope. It was about six hundred yards. Two tracers streaked across the field in the gathering darkness like swift angry bees. They plunged through the enemy machine-gunner's window, clearly marking it for the Marines below. Larger, angrier bees followed as the M60 on the ground floor sent a stream of tracers through the same window. The M60 pumped the distant room full of lead. Tracers trapped inside the concrete room ricocheted insanely in a weird kind of light-and-shadow show. My entries into the sniper's log for the rest of the night read simply: All Quiet.[2]

A few weeks later Corporal Rutter was again in action, suppressing fire from Shi'ite Amal militia against a sector of the Marine lines around Beirut International Airport. Rutter and his partner, Corporal James 'Rock' McGlynn, crossed the runway to be briefed by the 1st Platoon of Charlie Company – the target of the militia's small-arms and RPG fire. Journalist and ex-Marine sniper, Craig Roberts, described Rutter's mission:

With the morning sun, Rutter could see what lay in front of Charlie Company. To his immediate front, across a field, was a sign on a building proclaiming it the 'Danielle Café'. The café stood on a road that went straight away from Rutter's position. He could see the street well, and noted quite a bit of activity around the small restaurant.

On the wall of a building across the street from the restaurant, facing the Marines, was a large poster of Ayatollah Khomeini. Rutter grinned, 'There's one bastard I'd like to drop.'

As the two snipers watched, Amal militiamen entered the café and returned with weapons. The Shi'ites then took positions in bunkers and behind roadblocks. By late afternoon the natives grew restless. A few sporadic bursts of gunfire sailed over the Marine positions, but the shooters stayed behind cover. By nightfall, all was again quiet.

With dawn, Amal men again repeated their routine of entering the café to pick up their weapons. When they came out, they came out shooting. While most of the Shi'ites had reached their bunkers before the shooting began, a few were still left caught in the open on the street.

'I'll take the first one,' said Rock, drawing a bead on a moving figure carrying a Kalashnikov and wearing a green uniform and helmet liner. 'Must be a LAF [Lebanese Armed Forces] deserter. Still has his uniform.'

The M40A1 recoiled with the shot. Rutter saw the green helmet liner leap into the air, tumbling as it fell to the street. The man's head twisted grotesquely as brain matter erupted in a pink puff. The Shi'ite collapsed to the street.

The two men who had accompanied him looked down briefly, horror on their faces, then ran around the corner to safety — abandoning their comrade where he lay.

Two minutes later, a little red foreign car pulled around the corner and stopped. Two men jumped out of the back while the driver stayed behind the wheel, gunning the motor. The two Shi'ites ran to the fallen soldier and began dragging him towards the car. Another jumped out of the vehicle and began spraying bullets toward the Marine positions.

Rutter had already taken a bead on one of the men carrying the casualty. He let the cross-hairs settle and began to squeeze the trigger. Four hundred and twenty-five metres. A long range for an AK. An easy shot for a Marine sniper. The target flew backwards into the side of the car, his chest centred by the single bullet.

The second Shi'ite who had been dragging the dead man dove toward the safety of the car. Just then, another Marine sniper located nearby, Corporal Frank Roberts, shot through the driver's side of the windshield. The glass exploded just as the driver threw the car into reverse and began smoking the tyres. The little red car disappeared backwards around a corner. Two bodies lay in the street.

An old lady walked out of the building across from the Danielle Café and stared. She began yelling and pointing. The poster of the Ayatollah stared back at Rutter near the old lady. Rutter and McGlynn couldn't stand the temptation any longer. As people began running around the streets in confusion and the old lady waved her arms and yelled, the two Marines placed their sights right between Khomeini's eyes. Concrete dust exploded and the paper of the poster gapped wide as two 173-grain bullets found their mark. It was a message: From America, with love — asshole!

Including two shots fired into the poster, only five rounds were fired by the Marine snipers. Each man had struck exactly what he was aiming at.[3]

Immediately before the US Marines took up their positions in Beirut, the Israeli Defence Force had been pounding the city to destroy PLO influence in Lebanon. Chuck Kramer, an American who had played a part in training Israeli snipers, was despatched into the Lebanon to help in the battle against the PLO. This is his description of Israeli snipers in action in the ruins of Beirut:

> I was assigned an area south of the hippodrome which was right opposite the French Ambassador's house and had a lot of PLO infiltration. We had to make this area too hot for them to go into. The idea was to push them as far west into Beirut as we possibly could while the artillery just shot down buildings. I had the largest sniper force of the entire Israeli Army at that point. The numbers at the end of the war indicated there were forty-seven snipers employed in the entire action involving more than 100,000 troops. I had over twenty [of the snipers] under one command.
>
> The word [of our presence] had gotten out to the other side. Word spread like wildfire and these areas did quiet down immediately because they knew somebody was looking for them. They realized there was an organized group of Israeli snipers looking for them and this didn't sit too well.
>
> During this whole operation I would be in contact by radio with these guys and a confirmed kill [witnessed by two men] was recorded as 'one down'. I'd record the range for my own records. Most of the good hits were at extreme long range and most of the kills were made at 600–800 metres. [The PLO] were getting very wary by then. We spotted a guy on a motor scooter who seemed totally clean but who had weapons stashed in five or six compartments.
>
> I found to my surprise the PLO had very poor snipers, using equipment almost six or seven years old in comparison with modern Soviet equipment. They worked alone with no sort of training. They just went out to shoot as many Israeli

soldiers as they could. I found out – even as bad as they [the PLO snipers] were – the reaction of the Israeli soldier to being fired on by a sniper was terror. It was terrifying to feel you were under a sniper's scope. Units were stalled for hours while the commander was screaming for air support to take out a guy firing a plain Kalatch [AK-47 Kalashnikov] with no scope at 150 metres right in to his position. I found the Syrians had good snipers. They had good equipment and good training.

Most of the fighting was done in the first three days. The other twenty-six or so days there were countless cease-fires that were constantly violated by the PLO. That would cause the fighting to flare up again. In the first three days, we had sixteen [kills] verified, all at extended ranges, thirty-two probables and countless war stories.

The closest shot was 150 metres – a total fluke with a sniper rifle – where one of the snipers lined up on this guy and found to his horror he didn't have a round in the chamber. He pulled the bolt back on his M14 and eased it in and the bolt didn't lock. This happens with an M14. He lined up, squeezed and click. The guy he was shooting at heard the click and ducked back inside a building. The sniper jacked the round cursing in Hebrew, chambered another and waited. An Israeli air strike came over Beirut and these guys [the PLO] ran out like kids to watch the airplanes. That's when he shot him.[4]

The Iraqi invasion of Kuwait in August 1990 prompted a massive response from the West. When Saddam Hussein refused all demands to leave Kuwait, the scene was set for military action. The land campaign was preceded by a long and detailed air bombardment of Iraq, which destroyed communication systems, strategic industries and much of the Iraqi armed forces' will to fight. While the air war was under way, Allied patrols probed the Iraqi defences, and it was in these operations that snipers were employed to good effect.

In addition to their standard M40A1 rifles, the US Marines received a shipment of .50in calibre Barrett sniper rifles just before the onset of hostilities, and the desert proved an ideal testing ground for this heavy long-range weapon. In a number of instances, the Barrett was used to knock out light armoured vehicles, supplementing conventional anti-tank rockets. Hits at ranges in excess of 2000 metres were recorded on occasion.

Chris Hedges, a reporter on the *New York Times*, described the feelings and views of Marine snipers on the eve of the assault to reclaim Kuwait:

'When a sniper pulls the trigger, he can see the expression on a man's face when the bullet hits,' said Sgt Mark E. Anderson, chief scout of the platoon of Marine snipers and scouts breaking camp and getting ready to move at dawn the next day to the front lines along the Kuwait border.

The grand strategy of battle meant little to Staff Sgt Douglas A. Luebke, the non-commissioned chief of the platoon, and his men – other than that for the first time they will be called upon to go into combat, to do the job many have trained for years to carry out. They will crawl unseen within yards of enemy lines and, firing three shots from an exposed position, take out enemy soldiers one by one.

'It is the art of killing,' Sergeant Luebke said. 'We have to be perfect.'

The snipers said they were trying to avoid thinking of their Iraqi opponents as men with families and children. Several said the reputed maltreatment of American prisoners of war had steeled them for the task. 'I try not to think about the other man's personal life,' Sergeant Anderson said. 'I concentrate on him being the enemy. If I were to give him sympathy, I don't think I would be effective.'

Firearms are hardly new to most snipers. Nearly all the Marines in the platoon come from the country, where they grew up with weapons. They talk about stalking an animal and a man in the same breath. Most are passionate trap and skeet shooters. Their sentences are punctuated by the squirt of chewing tobacco and the inserting of profanities as adjectives.

The bonding of the unit, which will act as the eyes and ears of the battalion from forward positions, is tight. Platoon members are disdainful even of fellow Marines, who in turn chafe at the snipers' arrogance.

'We are as cocky as hell,' Sergeant Luebke said. 'It is the image we portray, because we know we are the best in the battalion.'

'You put me anywhere and I can put it in the head from 800 yards,' Sergeant Barrett said.

The work of the forward scouts and snipers is normally done at night, when two-man teams creep slowly toward enemy lines, dressed in camel-coloured camouflage smocks and wearing paint on their faces. The scouts and snipers call in artillery and aerial fire and try to kill enemy soldiers who operate heavy weapons like machine-guns and anti-aircraft weapons, usually at dawn or dusk.[5]

The success of the Marines in their sniper tasks can be seen in the figures of the 1st Marine Division, which secured thirty-nine kills for no losses, eleven of them at ranges over 800 yards.

In December 1992, US Marines spearheaded the arrival of a larger American military force in Somalia, aimed at separating the feuding warlords who had plunged the country into chaos. As in the Lebanon, US forces were severely restricted in their rules of engagement, and the Somalis were skilful in exploiting these restrictions. An extract from a letter sent home by a US sniper reflects some of the problems faced by the Americans, subjected to the full glare of media interest:

I'm sure you've read the wonderful story that is floating around the US from an AP reporter here. The Somalis here are well known for fabricating stories for the Press after we engage. In one incident I was accused of killing twelve people with one bullet.

On our last engagement we took out a man on top of a truck armed with an RPG [Rocket-Propelled Grenade]. The Somalis said we had killed a pregnant woman. An Army CID team was flown in from Washington the next day. A forensic expert examined the so-called victim and determined she had been hit by a moving vehicle and not by a .50 cal. bullet. By the time this was discovered the reporters had already written their stories that stated we were killing innocent women and children with hunting rifles and .50 calibre anti-armour weapons.[6]

Despite what the Marines saw as antagonism from the media and overly restrictive rules of engagement, snipers were again able to prove their effectiveness. US Marine sniper Lance-Corporal Andrew Lopez had been instructed at the 1st Marine Division Scout Sniper School in some of the more unusual sniper techniques, one of which included deliberately ricocheting a bullet on to an otherwise hidden target:

There was a wall alongside his target's hidden position, and Lopez decided to try to bounce a round off the wall and into or close to the enemy. Walls are not that easy to judge. A bullet tends to follow along a wall and does not come off the way a pool ball might angle from a table bumper.

At Lopez's shot there was movement and agitation in the enemy position. A little later, a patrol picked up Lopez's wounded opponent and gave him treatment. As a matter of interest they notified L-Cpl Lopez that his target was in custody, and Lopez went down to have a look.

Lopez's ricochet had worked well. Not a clean head hit, but a tumbling bullet penetration through the nose. In one side and out the other. Lopez warned the thoroughly cowed Somali that he had been merciful, but if he came again it would be his last appearance on earth.[7]

A more sustained piece of sniping, in which the Marines were able to get to grips with a Somali war band, is described in a narrative gleaned from interviews with several of the snipers involved:

> In Mogadishu, Somalia, a United States Army 10th Mountain Division HQ and Somali warlord [Mohammed Farrah] Aidid's tank and equipment park lay perpendicular to each other separated by 600 yards of open ground. The Somalis had been notified that they would have to surrender their heavy weapons on the morning of the next day, 7 January 1993.
>
> About 2130 hours on the 6th, Marines stationed at the stadium received a call for help from concerned 10th Mountain officers stating that Somalis were threatening them with tank cannon and some mobile four-barrelled anti-aircraft machine-guns. The soldiers claimed they were being intermittently fired upon by small arms. Puzzled, the Marines asked why the 10th did not fire back? The answer was that the HQ troops did not have the appropriate weapons. The soldiers had only M16s and machine-guns.
>
> Colonel Clent loaded three or four hundred Marines into their vehicles and moved them into position in and around the Somali and 10th Mountain buildings. Five scout snipers arrived and chose positions on top of the 10th's HQ building. It rained during the night, and conditions were unpleasant, but there was no fire from the Somalis. Morning light disclosed two M48 Somali tanks, two APCs, and two 220mm towed howitzers with their main guns pointing at the 10th Mountain HQ. There were also a pair of [four-barrelled] Soviet ZSU-23-4s.
>
> At 0625 hours loudspeakers began ordering the Somalis to surrender, and [US] Cobra gunships circled the area. Somalis rushed for their tanks, and the snipers received orders to take them out. At 600 yards, Marine Sgt Coughlin made the first shot with his Barrett .50 rifle. A Somali had climbed on to a ZSU four-barrel and was preparing to fire.

Coughlin's bullet killed him. The second anti-aircraft machine-gun opened up on the Marine snipers, chewing at the buildings, but Marine Sgt C. Johnson killed the gunner with one shot from his M40A1.

Of the Somali tankers attempting to gain their tanks, none made it. Sgt Johnson shot a rifleman who stood beside a tank and attempted to place fire on the HQ building. Marine Lance-Corporal Andrew Lopez, armed with an M40A1, killed three Somalis attempting to climb into tanks.

Cobra gunships placed at least one TOW missile into each Somali building, inflicting immense casualties, discouraging most from attempting to reach the tanks. A few enemy, foolishly, chose to fight.

A Somali was spotted by an observer, Cpl Green, and reported running for a tank. At 600 yards Sgt Johnson dropped him with a single round. The Somali soldier appeared to be shot in the hip and was dragged or crawled away before the firefight ended. Shooting was dying down when a hopeful Somali stuck an RPD [light machine-gun] out of a window and prepared to fire. The three Marine snipers manning rifles saw him almost simultaneously. All instantly fired. Later examination showed the body struck by one .50 calibre bullet and two calibre .30s.

No Somalis were successful in entering or operating their tanks, their APCs, or their towed artillery pieces. Later inspection showed all the Somalis' heavy guns to have rounds in their barrels, requiring only a pull to fire into the fragile 10th Mountain Division HQ building.[8]

In the war that followed the break-up of Yugoslavia in 1991, soldiers of United Nations were repeatedly fired upon by Croat, Serb or Muslim gunmen. Often, they were unable to return fire, which only encouraged the gunmen to continue their sniping. Occasionally, exasperated UN troops were able to retaliate. In one instance, early in January 1993, a relief convoy, escorted by the British Army, was shot at three times in a row by a Serb sniper;

on the third occasion, three Scimitar 'light tanks' rapidly returned fire, expending 17 30mm cannon rounds and 125 machine-gun rounds. Serb sniper fire ceased.

Despite a long hunting tradition and experience in rifle shooting, the Yugoslavs had little knowledge of military sniping beyond the most basic level. As a consequence – especially when the war between Serbia and Croatia increased in intensity – there was a growing demand for foreign sniper trainers. John L. Hogan, an American sniper who had served in Vietnam and had subsequent experience as a military trainer, worked on the Croat side.

Based around the town of Vinkovci, Hogan organized the acquisition of suitable sniper rifles, which ranged from the ex-Yugoslav Army M76 sharpshooter rifle (based on the Soviet SVD), through civilian hunting rifles of various calibres to the Austrian Steyr SSG. His next task was to recruit suitable candidates. Having gathered thirty men, divided into fifteen two-man teams, Hogan decided to instruct just three teams directly. They, in turn, would become trainers of the other teams. Before he could begin, however, Hogan had to deal with an attitude problem among his recruits: 'Every Croatian knew in his heart he was born a dead shot. It was part of being a man. One afternoon with a breeze from 11 o'clock, I set up a target that they had to guess was 650 meters and let them have at it. Not a single hit was recorded on a two-by-four-foot target; their attitude changed.'[9]

Using the US Marine scout sniper manual as a training syllabus, Hogan set about transforming his recruits into snipers. Besides intensive range work, firing from 300 to 900 metres, the recruits were instructed in advanced fieldcraft, emphasizing movement, concealment and camouflage. This included the construction of ghillie suits by each man.

Hogan learned from Croat field intelligence that former-Yugoslav special forces had taken over a nearby village and were sniping at Croat troops in the area. Satisfied that his men had achieved a reasonable standard, he decided that a live field exercise would form ideal training in counter-sniper work. Just before dark, Hogan and his three teams infiltrated the deserted village, using a Croat patrol as cover. Hogan relates the course of their sniper mission:

> Most of the buildings had sustained hits, so there were plenty of holes and piles of rubble to hide in. During the night we built five hasty hides in the rubble; I put two teams in two of them with the strict instruction that they should fire no more than two shots before moving back to another hide. I took the third team about 200 meters to the second storey of a house that had a good view of the back of all five hides, so that we could provide security if things began to turn sour with the two working teams.
>
> By daybreak neither team had fired a shot, which was good because I wasn't looking forward to trying to navigate the village in the dark if they got into trouble. About 0700, the man downstairs came up to our floor whispering excitedly. When the translation came through, he was saying he had heard movement nearby; all three of us went into full alert. I kept the Steyr with me and followed him back downstairs, leaving one man upstairs with binoculars to watch the hides. We both readied grenades and hid behind piles of timber and brick to wait. Soon I heard the noise outside. There was clearly movement in the garden, but no voices.
>
> After a while, we heard pig noises and looked out. A large pig was tearing flesh off the corpse of an emaciated milk cow that had probably been hit the day before. I felt easier, but the Croatian was clearly worried about the pig and wanted to shoot it. The reason he was so worried was the fact that pigs turned carnivorous by hunger will quickly attack a man. Throughout the morning we watched other livestock and domestic animals moving around the village. My worst fear was that one of the pigs would sniff up a sniper team and we would have to blow our position to kill the beast.
>
> Aside from a mamma cat and three kittens, nothing

came close to the hides that afternoon, but around 1500 [hours] I heard a proper scuffle downstairs and a loud pig snort. I was on the way downstairs when I met my man coming up, wiping blood off his knife. 'Serbski sveinjetina,' he said – Serbian pork. He had patiently waited and cut the pig's throat as it came through a hole in the wall.

By the time the sun began to sink, we hadn't fired a shot, had only a dead pig to show for our time and the Croatian patrol would soon be coming to pick us up. It was almost dark. I'd pretty much given up and was back in the shadows shucking up my ghillie suit when the forward hides fired once each. I was on my belly with the binoculars when one of the hides fired a second time. Within a moment or so, two more shots sounded, but not from the hide – these came from our flank and slightly to our rear. It could be a Croatian, but aside from the patrol there weren't supposed to be any in this part of the village.

I grabbed my team-mate and moved as quickly as possible to the other side of the house. The entire top corner had been blown off on that side, so he aimed his rifle in the general direction of the sound while I scanned the area. Sure enough, there was another muzzle flash, up high in a group of houses about 200 metres away.

I heard my sniper set his double triggers and hoped that the other guy would take just one more shot. He did – in fact, he took two more, clearly firing a semi-automatic. No sooner had the second shot flashed when the .300 WinMag next to me fired once. My sniper didn't move out of position as he cycled a fresh cartridge into the chamber. I was proud of him. Even if he missed, he held well and was moving correctly – almost not at all. We watched for a while, but no other muzzle flashes appeared in that direction.

After linking up with our two teams and the Croatian patrol, we went to recon the house from where the semi-auto fire had come. The Croatians checked the house and motioned me in. There was a man in Yugo army camo lying on the floor with a silver-dollar-size exit hole in his spine, where his back met his neck. Turning him over, there was an entry hole just above the clavicle where the 180-grain Barnes X [bullet] had entered. Someone had obviously been with him because his rifle was gone. But the empty cases on the floor were 7.92×57mm with Cyrillic headstamps. This told me there was at least one Yugo-trained sniper who didn't know enough to acquire his target properly or to move after his second shot.

After we got out of the village, the two forward teams explained their three shots. They had engaged a Serb patrol coming out of the cornfields. Both teams had put one man down each at about 300 metres with their first shots. Then, when one of the Serbs' buddies had come out from cover to check the two on the ground, one of my guys had laid him down with another shot. Four rounds fired – three hits and one confirmed KIA.[10]

'Skippy' Hampstead, an experienced ex-soldier, was involved in sniping operations in Croatia and Bosnia. Trained in the Australian Army, Hampstead worked alongside a former British Army sniper, call-signed Rose, fighting with Croats against the Serbs. In many areas, unless there was a major offensive, operations tended to be static, each side facing the other across no man's land. During one such lull in the fighting, Hampstead acted as heavy weapons back-up to his partner:

At night we'd go out and find ourselves somewhere which had good protection, just on the front line overlooking the enemy positions which were across a valley. We knew all the ranges and where their positions were – about 400–500 metres away – and we zeroed in our weapons to that range. Rose had a Yugoslav M76 with a night sight and I had an MG42 machine-gun, with every third round a tracer round.

He'd locate a target with his night sight and fire, and I'd be firing short bursts with the MG42 and in a couple of seconds I'd be on target.

On one particular occasion he dropped a Serb – he'd probably been having a piss or something – and he was left there screaming. Then all his mates came running out to help him and I got them with the machine-gun. I gave them two or three bursts before I shut up and we moved away. With all that tracer it was very graphic, and everything was lit up. I couldn't see the results, but Rose could spot the impacts through his image intensifier. These night actions worked on more than one occasion.[11]

The use of a machine-gun alongside a conventional sniper had proved an effective combination in many previous conflicts. A later development was the employment of lightweight anti-armour weapons as an adjunct to sniping. While operating in an urban environment, Hampstead deployed a 66mm rocket launcher against troops of the Serbian JNA (Yugoslav National Army):

I remember we were in these buildings opposite the JNA barracks, which were across the river from us. For some reason we had with us a 66mm one-shot, anti-tank rocket launcher. We spotted three Serb soldiers at one of the windows in the barracks, and we debated whether to use the rocket launcher. The back-blast in a building would be phenomenal, but in the end we went ahead. Rose deliberately attracted their attention, waving at them a few windows away from where I was. They were curious but they didn't try to engage us. Then Rose yelled, 'Go on! Fire at them!' I fired. The ceiling behind seemed to disintegrate and there was smashed glass and smoke everywhere. We were deafened and half-blinded. But when we looked over at the barracks, there was just a hole and all the walls around the window had turned black. We very quickly got out of there. It was OTT sniping but it worked.[12]

Having fought the Serbs in Croatia, Hampstead and Rose were transferred to Bosnia, where they were engaged against Serb forces for control of the city of Mostar. There, they were involved in a counter-sniper action:

When we were in Mostar, Rose always used to emphasize to me, 'Watch for other snipers. When you're looking for targets, remember other counter-snipers are going to be around.' You had to follow the classic things like looking for shadow, silhouette or glint. You really had to concentrate.

In Mostar we were up against this Serb sniper. He was really into terror – kids, women, he didn't give a fuck. He was a murderer. We really had to nail this bloke, give him a third eye. I personally wanted to do the hit, especially as he had wasted a mate of mine. We knew where his shots had come from; we'd witnessed a few misses. There was one particular part of the road – heavily used by civilian traffic – which he regularly fired at. We roughly worked out the distance and the position from which he was firing.

It took us a while to get him, but one day we saw him in the rafters of a house on a hillside. He was standing well back like a proper Yugoslav Army trained sniper, but we saw his white face, clean shaven, firing through the beams, relying on the shadow to hide him. On this particular occasion, his white face bobbing around gave him away. We waited for about 10 minutes until he showed himself properly, bringing his rifle up and taking aim. Rose took him with one shot.

It was a good shot against a small target about 350 to 400 metres away. We couldn't afford to miss. One missed shot and he'd fuck off and go somewhere else, and we would have lost the whole thing. People had looked for him before and got nowhere – or ended up casualties. It was great when we got him. Our commander heard about it, and I think it even got into the papers in Mostar.[13]

Killing a man who had terrorized the local population posed few moral problems for a sniper, but in other instances, shooting a man in cold blood was more difficult. Hampstead explains some of the feelings going through his mind as he viewed his target:

> When you're looking down the scope at someone, I sometimes got this feeling that I was the hand of God. It was up to me whether this bloke lived or died. He had no idea that you were there, and if you hadn't dropped the hammer on him, then he'd never have known. But if you did, then that's one life less, with brothers and sisters, mothers and fathers, grandparents, cousins, uncles all affected by this one action, which was not always necessary. After all, it was not your life being threatened.
>
> Sniping is a very intense personal and emotional battle. You have to psych yourself up to go out and kill another man. When you sit there watching your target for ten minutes, sometimes you get a sort of personal affinity with the person. To be successful you had to be prepared to overlook the fact that you were taking a human life. Although they were soldiers in uniform, and they were aggressors who shouldn't have been there, it could still be a problem. Sometimes you could see them laughing, carrying on, scratching their arse or lighting a cigarette.
>
> Once I saw two Serbs trying to light a cigarette with a lighter that didn't work properly. I didn't know which one to shoot. I decided in the end I'd shoot the one who lit the cigarette first. How random is that? And yet it happened. Somewhere there is someone saying that this guy who was standing next to me lit a cigarette and his head exploded. That guy will never know the 'what' or the 'why' of how he survived.[14]

A soldier from the British Parachute Regiment, Lee Marvin was another volunteer fighting for the Croats against the Serbs. While waiting to join up with his main Croat unit, he commanded an

assault against a Serb stronghold with a group of Croat soldiers. Marvin divided his force into a reconnaissance section and a heavy weapons section, with a machine-gun on the right flank. Marvin and another machine-gun team would provide support on the left flank for the main attack. Armed with a Yugoslav M76 sniper rifle – a copy of the Soviet SVD Dragunov – Marvin and his support team carefully moved into position to eliminate the Serb machine-gun platoon sited in a forward trench. The unsuspecting Serbs were easily visible to the Croat force. Marvin's sniping took place at very close range, but the action illustrated the need for quick follow-up sniper fire and steady nerves:

I slowly made my way forward with the Dragunov over my shoulder and my AK-47 in my hand, just in case I was seen going in. All was very quiet, and all I could hear was the Serbs laughing and joking. There didn't seem to be many of them either, but that was outside; inside [the main position] there could be any number, but they didn't really stand much chance because once the attack commenced anyone who tried to leave the building would have it blown up on top of them.

I told the lads to hold fast until they heard at least three shots coming from me, and then they could start their advance. Our plan was a simple one, but effective, if the lads didn't fuck it up, that is. I and my section would take out the three main bunkers and then the MG section would give covering fire in the direction of the houses, so that the Serbs would keep their heads down, while the Croats advanced on their position.

I moved forward as far as I could, almost to the end of the trees that were covering me. I was cammed up the best I had ever been, for this time it was the most important as I'd never been so close to someone who would definitely kill me if he knew I was there. I had been taught well and the fact that I was still alive was proof of the matter. The feeling I had was indescribable, a mixture of fright and excitement. I carefully put the AK-47 on to the ground and took the Dragunov from my shoulder. I drew the weapon into my shoulder and put the sling around my right arm to get maximum control of the weapon and took the first Serb sentry into my sights. I then moved the sights at the second and then the third, then back to the second and finally back to the first. I continued this process for a minute or two to make sure that I had definitely got their positions and then for the final time I took the first sentry into my sights.

I released my hand from the grip and took off the safety catch and then replaced my hand very carefully so as not to disturb my aim or knock my concentration. I needed all of it to get those three sat in front of me: the first was only about 75 feet from me, the second around 110 feet and the third about 200 feet away. I was more than confident that I could make the shots count, the only trouble being after I had hit the first one, the other two were not likely to remain where they were and do nothing; they would obviously pick up their weapons and open fire on my position.

I started to control my breathing, so as I inhaled the sights moved down over the target and as I exhaled the sights would slowly move back again. I did this a few times and finally as I exhaled slowly, gently squeezed the trigger. From that moment everything seemed to go in slow motion; of course, it wasn't really. The first sentry fell to the ground. I took aim on the second and he was hit before he had a chance to move. Then the third; but as he had moved, caught it in the shoulder blade, and although he went down from the force of the shot he began to get up and grabbed his weapon. As he did so, I pulled the trigger one last time. I think I hit him in the side of the neck. As he fell to the ground, all hell broke loose, with both the machine-guns opening up.[15]

The attack was successful, and the Serbs suffered heavy casualties, although the Croats lost men to enemy mines. While many US and British snipers would disparage the M76/Dragunov for its inability to operate effectively at ranges beyond 600 metres, in this instance, the rifle's semi-automatic capability made it an ideal short-range sniper weapon.

The Bosnian capital Sarajevo proved to be almost perfect for indiscriminate sniping, especially as the Serbs held most of the high ground surrounding the city, and delighted in targeting Bosnian civilians in the streets below. In an attempt to eliminate these Serb gunmen, the Bosnian government formed counter-sniper units. One unit was visited by Robert Mackenzie, a contributing editor to *Soldier of Fortune* magazine:

> The Bosnian counter-snipers knew their jobs. Their hides were well back from windows or shell holes, and they had easy routes to dozens of alternates. Since Serbian counter-fire often took the form of a tank round, 'shoot and scoot' was one of their most closely followed operating principles. With every two or three shooters was an observer equipped with binoculars or 40×42 spotting scopes, who also carried maps and sketches of scores of known Serbian positions and bunkers.
>
> Climbing many, many storeys up the parliament building, I had a firsthand look at the Bosnians' biggest problem in trying to suppress enemy snipers: on the hillside opposite were literally thousands of empty windows, and thousands more trees, bushes and piles of debris, each one a potential firing position. One of the worst areas was an overgrown cemetery about 300 metres away where each crypt and tombstone could hide a rifleman. Clearly, those Serbs were proficient in their tactics and use of ground.
>
> Ironically, some of the Serbian sharpshooters had served in the same police counter-terrorist unit as their Bosnian counterparts, and they were all known by name to each other. Perhaps the most discussed was a former shooting instructor who had trained some of the people now trying to shoot him. An Olympic-medallist marksman, he was reputed

to be very capricious in his choice of targets: one week only women, the next week only journalists, the following week firemen, etc. His former students were very much in dread of his cross-hairs, as he reportedly never, ever missed. If he dropped the hammer on you, you got more than your ticket punched.[16]

Bosnian civilians were not the only targets for snipers. UN peacekeepers faced attacks from all three warring sides – Serbs, Croats and Muslims. The French were responsible for holding the line in Sarajevo, and developed their own counter-sniper units, although they faced extra problems. Some of these were outlined by Colonel Guy de Haynin de Bry, the commanding officer of a French paratrooper battalion deployed in Sarajevo. In an interview with Richard Lucas in February 1995, de Haynin de Bry said:

Our target is not an easy one. The sniper can choose his time and place. He can hit and run, fire once or twice and be out before we have time to fire back. When he does work from a fixed position he is usually well dug in, shooting out of improvised but effective bunkers much like those we encountered in Beirut.

From a military point of view, a search-and-destroy operation is what's called for and I have the men capable of carrying out that type of mission. But we simply can't go in and take them out. They are operating in a zone which is out of our jurisdiction. We are here, in Sarajevo, essentially to implement a policy of sniper dissuasion. If it is necessary physically to eliminate the sniper in order to protect lives and make this policy work, my men have orders to eliminate the sniper and remove the threat.

What complicates the situation is the fact that the two opposing sides are in extreme proximity, sometimes only twenty or thirty metres apart, and they are equipped with the same weapons. All this makes identification that more difficult.[17]

De Haynin de Bry's men used mobile firebases to counter the sniper threat, Armoured Personnel Carriers (APCs) with a five-man counter-sniper team aboard. They were equipped with the vehicle's 20mm cannon and co-axial 7.62mm machine-gun, plus a .50 McMillan heavy sniper rifle and a 7.62mm FRF2 sniper rifle. Richard Lucas describes the APCs in use:

When I arrived at Sniper Alley during my last trip, I could see five of the white UN APCs spaced out about 150 metres apart, their guns pointed in the direction of the Serb-held buildings on the other side of the river. As I approached one of the APCs I could see a trooper inside scanning the area with binoculars. As he called out fire co-ordinates, another trooper followed with the scope of the FRF2 sniper rifle and verified each one. A sketched-out panoramic view of the area was taped to the bulkhead in front of them.

I climbed into the back of the APC and the team sergeant explained the activity. The mapped-out co-ordinates on the sketch code-named each building in the area and designated each window or other possible sniper site with a letter and a number. In this way the spotters and the fire teams had the same reference, and any activity could be radioed in and then located and followed by the other teams in that sector.

'We run through the sector co-ordinates a few times a day,' explained the sergeant, 'especially whenever someone takes over surveillance, so that everyone is familiar with the area. Also we've got about a dozen possible sniper positions in our sector which we keep under observation. After that, except for making sure the weapons are ready to function and the periodic radio checks, it's a waiting game.'[18]

Corporal Chevardez, a veteran of conflicts in Chad and Zaïre, who manned the .50 McMillan, described a recent counter-sniper action:

One of the spotters picked up movement in our sector and radioed for co-ordinates. The [enemy] sniper was setting up in one of the white block apartments across from us, about 350 metres away. I was already on target when I heard the shots, and the smoke gave me the visual confirmation I needed. I had time to get off two rounds, both explosive, before the 20mm swung into action and sent in a volley. That took a good piece out of the wall and that was it. I think I got him. Anyway, it's been quiet from up there ever since.[19]

The extraordinary high level of opportunistic sniping that took place in the fighting for the former Yugoslavia was once again confirmation of the importance of sniping in modern warfare. Those armed forces which deploy trained snipers will always have a tactical edge over those that neglect this most vital of military disciplines.

At the Receiving End

This book has been concerned with the sniping from the sniper's perspective. But what of his victim? Death, of course, curtails any comment, but sufficient numbers of casualties have survived to tell their tale. The writer George Orwell was shot in the neck by a sniper while serving in the Republican forces during the Spanish Civil War (1936–39). In his book *Homage to Catalonia*, Orwell wrote at length of the sensations he experienced when hit by the bullet. Despite the seriousness of the wound and some indifferent medical treatment, he survived to make a full recovery.

Orwell had been sent to the front to man a quiet sector of the line, and there he began to snipe the Fascist soldiers in the trenches opposite. Readily acknowledging that he was a poor shot, he had little success, but, more damning, his own carelessness made him vulnerable to enemy fire:

> It was at the corner of the parapet, at five o'clock in the morning. This was always a dangerous time, because we had the dawn at our backs, and if you stuck your head above the parapet it was clearly outlined against the sky. I was talking to the sentries preparatory to changing the guard. Suddenly, in the very middle of saying something, I felt – it is very hard to describe what I felt, though I remember it with the utmost vividness.
>
> Roughly speaking it was the sensation of being at the centre of an explosion. There seemed to be a loud bang and

a blinding flash of light all round me, and I felt a tremendous
shock – no pain, only a violent shock, such as you get from
an electric terminal; with it a sense of utter weakness, a
feeling of being stricken and shrivelled up to nothing. The
sandbags in front of me receded into immense distance. I
fancy you would feel much the same if you were struck by
lightning. I knew immediately that I was hit, but because of
the seeming bang and flash I thought it was a rifle nearby
that had gone off accidentally and shot me. All this happened
in a space of time much less than a second. The next moment
my knees crumpled up and I was falling, my head hitting
the ground with a violent bang which, to my relief, did not
hurt. I had a numb, dazed feeling, a consciousness of being
very badly hurt, but no pain in the ordinary sense.

The American sentry I had been talking to had started
forward. 'Gosh! Are you hit?' People gathered round. There
was the usual fuss – 'Lift him up! Where's he hit? Get his
shirt open!' etc., etc. The American called for a knife to cut
my shirt open. I knew that there was one in my pocket and
tried to get it out, but discovered that my right arm was
paralysed. Not being in pain, I felt a vague satisfaction. This
ought to please my wife, I thought; she had always wanted
me to be wounded, which would save me from being killed
when the great battle came. It was only now that it occurred
to me to wonder where I was hit, and how badly; I could feel
nothing, but I was conscious that the bullet had struck me
somewhere in the front of the body. When I tried to speak I
found that I had no voice, only a faint squeak, but at the
second attempt I managed to ask where I was hit. In the
throat, they said. Harry Webb, our stretcher-bearer, had
brought a bandage and one of the little bottles of alcohol
they gave us for field-dressings. As they lifted me up a lot of
blood poured out of my mouth, and I heard a Spaniard
behind me say that the bullet had gone clean through my
neck. I felt the alcohol, which at ordinary times would sting
like the devil, splash on to the wound as a pleasant coolness.

They laid me down again while somebody fetched a stretcher. As soon as I knew that the bullet had gone clean through my neck I took it for granted that I was done for. I had never heard of a man or animal getting a bullet through the middle of the neck and surviving it. The blood was dribbling out of the corner of my mouth. 'The artery's gone,' I thought. I wondered how long you last when your carotid artery is cut; not many minutes, presumably. Everything was very blurry. There must have been about two minutes during which I assumed that I was killed. And that too was interesting — I mean it is interesting to know what your thoughts would be at such a time. My first thought, conventionally enough, was for my wife. My second was a violent resentment at having to leave this world which, when all is said and done, suits me so well. I had time to feel this very vividly. The stupid mischance infuriated me. The meaninglessness of it! To be bumped off, not even in battle, but in this stale corner of the trenches, thanks to a moment's carelessness! I thought, too, of the man who had shot me — wondered what he was like, whether he was a Spaniard or a foreigner, whether he knew he had got me, and so forth. I could not feel any resentment against him. I reflected that as he was a Fascist I would have killed him if I could, but that if he had been taken prisoner and brought before me at this moment I would merely have congratulated him on his good shooting. It may be, though, that if you were really dying your thoughts would be quite different.

They had just got me on to the stretcher when my paralysed arm came to life and began hurting damnably. At the time I imagined I must have broken it in falling; but the pain reassured me, for I knew that your sensations do not become more acute when you are dying. I began to feel more normal and to be sorry for the four poor devils who were sweating and slithering with the stretcher on their shoulders. It was a mile and a half to the ambulance, and vile going, over lumpy, slippery tracks. I knew what a sweat it was, having helped to carry a wounded man down a day or two earlier. The leaves of the silver poplars which, in places, fringed our trenches brushed against my face; I thought what a good thing it was to be alive in a world where silver poplars grow. But all the while the pain in my arm was diabolical, making me swear and then try not to swear, because every time I breathed too hard the blood bubbled out of my mouth.

The doctor rebandaged the wound, gave me a shot of morphia, and sent me off to Sietamo. The hospitals at Sietamo were hurriedly constructed wooden huts where the wounded were, as a rule, only kept for a few hours before being sent on to Barbastro or Lerida. I was dopey from morphia but still in great pain, practically unable to move and swallowing blood constantly.[1]

Source Notes

Chapter One
The Long Rifle and the American Revolution

1. Harold L. Peterson, *The Book of the Gun* (Paul Hamlyn, 1963), p. 140.
2. George Washington, from Sparks, *The Writings of George Washington*, vol. VII, p. 461, quoted by Joe D. Huddleston, 'Timothy Murphy — Revolutionary Riflemen' in Craig Boddington, ed., *America: the men and their guns that made her great*, p. 8.
3. Charles Lee, quoted in Charles W. Sasser and Craig Roberts, *One Shot — One Kill* (Pocket Books, 1990), p. 240.
4. Charles Winthrop Sawyer, *Firearms in American History* (privately published, 1910), pp. 80–81.
5. Sawyer, op. cit., p. 84.
6. Lieutenant William Digby, *Journal*, quoted in Henry Steele Commager and Richard B. Morris, *The Spirit of the Seventy-Six* (Harper and Row, 1958), p. 580.
7. Captain E. Wakefield, *Recollections*, quoted in Commager and Morris, op. cit., p. 581.
8. Samuel Woodruff, quoted in Commager and Morris, op. cit., p. 593.
9. Digby, quoted in Commager and Morris, op. cit., p. 597.
10. Lieutenant-Colonel J.G. Simcoe, *Simcoe's Military Journal: A History of the Queen's Rangers* (Bartlett and Welford, 1844), p. 38.
11. Brigadier-General Simon Fraser, 'Memorandum relative to a

Company of Marksmen', quoted in Stephen G. Strach, 'A Memoir of Captain Alexander Fraser' in the *Journal of the Society of Army Historical Research*, Spring 1985, p. 96.

12. Hans Busk, 'Handbook for Hythe', quoted in Berkely R. Lewis, *Small Arms and Ammunition in the United States Service* (Smithsonian Institute, 1956), p. 90.

13. Colonel George Hanger, *A Letter to the Rt Hon. Lord Castlereagh* (1808), and *To All Sportsmen* (1814).

14. A Corporal of Riflemen [Captain Henry Beaufoy], *Scloppetaria, or Consideration of the Nature and Use of Rifled Barrel Guns* (1808), p. 29.

Chapter Two
Sharpshooting in Europe

1. A Corporal of Riflemen (Captain Henry Beaufoy), *Scloppetaria, or Consideration on the Nature and Use of Rifled Barrel Guns* (1808), p. 23.

2. Money, quoted in David Gates, *The British Light Infantry Arm c.1790–1815* (Batsford, 1987), p. 59.

3. William Surtees, *Twenty-Five Years in the Rifle Brigade* (Blackwood, 1833), pp. 289–90.

4. George Hanger, quoted in Gates, op. cit., p. 85.

5. Sir Henry Bunburry, quoted in Gates, op. cit., p. 66.

6. Beaufoy, op. cit., p. 22.

7. Surtees, op. cit., pp. 16–17.

8. Beaufoy, op. cit., p. 27.

9. Order book, Fifth Battalion, 60th Foot, quoted in Gates, op. cit., p. 96.

10. Beaufoy, quoted in Gates, op. cit., p. 82.

11. Surtees, op. cit., p. 47.

12. Edward Costello, *The Peninsular and Waterloo Campaigns*, ed. by Anthony Brett-James (Longmans, 1967), p. 68.

13. Costello, op. cit., pp. 91–2.

14. Lieutenant-Colonel J. Leach, *Rough Sketches of the Life of an Old Soldier* (Longman, 1831), p. 255.

15. John Harris, *The Recollections of Rifleman Harris*, ed. by Christopher Hibbert (Leo Cooper, 1970), pp. 26–7.
16. Harris, op. cit., p. 38.
17. Private Wheeler, *The Letters of Private Wheeler*, ed. by Captain B.H. Liddell Hart (Michael Joseph, 1951).
18. Major Reginald Hargreaves, 'The Lonely Art' in the *Marine Corps Gazette*, December 1954, p. 72.
19. Lieutenant-General Sir George MacMunn, quoted in Arthur Swinson, *North-West Frontier* (Hutchinson, 1967), pp. 247–8.
20. Hargreaves, op. cit., pp. 72–3.

Chapter Three
The American Civil War

1. New York Post, 4 June 1861, quoted in Roy M. Marcot, *Civil War Chief of Sharpshooters: Hiram Berdan – Military Commander and Firearms Inventor* (Northwood Heritage Press, 1989), p. 22.
2. William F. Ripley, *Vermont Riflemen in the War for the Union – A History of Company F, First United States Sharpshooters* (Tuttle & Company, 1883), quoted in Marcot, op. cit., p. 24.
3. Rudolf Aschmann, *Memoirs of a Swiss Officer in the American Civil War*, ed. and int. by Heinz K. Meier (Herbert Lang, 1972), p. 30.
4. Charles Stevens, *Berdan's United States Sharpshooters in the Army of the Potomac – 1861 to 1865* (Morningside Bookshop, 1984), quoted in Marcot, op. cit., p. 43.
5. Aschmann, op. cit., pp. 30–31.
6. Ripley, op. cit., quoted in Marcot, op. cit., p. 122.
7. Stevens, op. cit., quoted in Marcot, op. cit., p. 133.
8. Aschmann, op. cit., p. 58.
9. Aschmann. op. cit., p. 63.
10. *Civil War Diary of First Sergeant Wyman S. White* (unpublished), quoted in Marcot, op. cit., p. 122.
11. Ed Thompson, *History of the Orphan Brigade* (Press & Bindery of Chas. T. Dearing, 1898), quoted in John Anderson Morrow, *The Confederate Whitworth Sharpshooters* (privately published, 1989), p. 35.

12. Sam R. Watkins, *Company Aytch, A Side Show of the Big Show* (Cumberland Presbyterian Publishing House, 1882), quoted in Morrow, op. cit., p. 35.

13. William C. Davis, *The Orphan Brigade: The Kentucky Confederates Who Couldn't Go Home* (Doubleday, 1980), p. 216.

14. John West, quoted in Morrow, op. cit., pp. 42–3.

15. Watkins, op. cit., p. 145, quoted in Morrow, op. cit., p. 41.

16. John W. Green, *Johnny Green of the Orphan Brigade*, ed. A.D. Kirwan (University of Kentucky Press, 1956), pp. 176–7, quoted in Morrow, op. cit., p. 65.

17. F.S. Harris, *Confederate Veteran*, vol. IV, p. 73, quoted in Morrow, op. cit., p. 73.

18. John N. Edwards, *Shelby and His Men*, quoted in Morrow, op. cit., p. 34.

19. Davis, op. cit., pp. 226–7.

20. Dan Flores, 'Sharpshooters in the Civil War' in *Gun Digest*, 31st ed., 1977, p. 13.

21. Aschmann, op. cit., pp. 154–5.

22. Frank Vizetelly, 'The War in America: Confederate Sharpshooters Firing on a Federal Supply-Train on the Tennessee River' in *The Illustrated London News*, 5 December 1863, p. 574.

23. Rice C. Bull, *The Civil War Diary of Rice C. Bull, 123rd New York Volunteer Infantry*, ed. Jack Bauer (Presidio Press, 1977), p. 136.

24. Aschmann, op. cit., pp. 58–9.

25. Ripley, op. cit., quoted in Marcot, p. 47.

26. S.A. Ashe, *Confederate Veteran*, vol. XXXV, pp. 254–5, quoted in Morrow, op. cit., p. 55.

27. Thomas B. Brookes, in *The War of Rebellion: A Compilation of Official Records of the Union and Confederate Armies*, Series I, vol. XXVIII, Part I (US Govt, 1880–1901), p. 264, quoted in Morrow, op. cit., pp. 56–7.

28. John S. Jackman, *Diary of a Confederate Soldier: John S. Jackman of the Orphan Brigade*, ed. William C. Davis (University of South Carolina Press, 1990), p. 129.

29. Jackman, op. cit., p. 132.

30. Aschmann, op. cit., pp. 151–2.

31. Aschmann, op. cit., pp. 170–71.

32. J.W. Minnich, *Confederate Veteran*, vol. XXX, pp. 294–6, quoted in Morrow, op. cit., pp. 64–5.

33. Stevens, op. cit., quoted in Marcot, op. cit., p. 42.

34. Quoted in Wiley Sword, *Sharpshooter: Hiram Berdan, his famous Sharpshooters and their Sharps Rifles* (Andrew Mowbray Inc., 1988), p. 47.

35. James K. Hosmer, from William A. Spedale, *Where Bugles Called and Rifles Gleamed* (Land and Land Publishing Division, 1986), pp. 135–6, quoted in Morrow, op. cit., pp. 76–7.

36. Quoted in Sword, op. cit., pp. 58–9.

37. Jackman, op. cit., p. 132.

38. *Confederate Veteran*, vol. XVI, p. 172, quoted in Morrow, op. cit., p. 79.

Chapter Four
The First World War

1. Major H. Hesketh-Prichard, *Sniping in France: With Notes on the Scientific Training of Scouts, Observers and Snipers* (Hutchinson, n.d.), pp. 25–6.

2. Hesketh-Prichard, op. cit., p. 29.

3. Frank Richards, *Old Soldiers Never Die* (Anthony Mott, 1983), p. 62.

4. Harold Stainton, 'A Personal Narrative of the War', Department of Documents, Imperial War Museum.

5. Richards, op. cit., p 42.

6. Major F.M. Crum, *With Riflemen, Scouts and Snipers* (privately published, 1921), pp. 78–9.

7. *Student and Sniper Sergeant: A Memoir of J.K. Forbes*, ed. William Taylor and Peter Diak (Hodder and Stoughton, 1916), pp. 116–17.

8. Forbes, op. cit., pp. 110–11.

9. Quoted in *The History of the Seventh (Service) Battalion of the Royal Sussex Regiment 1914–19*, ed. Owen Rutter (Times Publishing Company, 1934), pp. 25–6.

10. Crum, op. cit., pp. 2–3.

11. Hesketh-Prichard, op. cit., p. 36.

12. Herbert W. McBride, *A Rifleman Went To War* (Small-Arms Technical Publishing, 1935), p. 300.

13. Forbes, op. cit., p. 138.

14. Forbes, op. cit., pp. 136–7.

15. Hesketh-Prichard, op. cit., p. 75.

16. Crum, op. cit., pp. 112–13.

17. F.A.J. Taylor, *The Bottom of the Barrel* (Regency Press, 1928), p. 118.

18. Crum, op. cit., pp. 82–3.

19. Frederick Sleath, *Sniper Jackson* (Herbert Jenkins, 1919), pp. 179–81.

20. Ralph Hodder-Williams, *Princess Patricia's Canadian Light Infantry* (Hodder and Stoughton, 1923), vol. I, p. 25.

21. Hesketh-Prichard, op. cit., pp. 44–5.

22. McBride, op. cit., p. 82.

23. Ion L. Idriess, *The Desert Column* (Angus and Robertson, 1932), pp. 52–3.

24. McBride, op. cit., p. 95.

25. Sleath, op. cit., pp 184–5.

26. McBride, op. cit., p. 89.

27. Hesketh-Prichard, op. cit., pp. 99–100.

28. McBride, op. cit., pp. 160–61.

29. Thomas Baker, audio interview, Department of Sound Records, Imperial War Museum.

30. Idriess, op. cit., pp. 43–5.

31. Baker, op. cit.

32. Ion L. Idriess, *The Australian Guerrilla – Book II: Sniping* (Angus and Robertson, 1942), pp. 89–90.

33. Quoted in Tony Ashworth, *Trench Warfare 1914–18 – the Live and Let System* (Macmillan, 1980), p. 109.

34. Ion L. Idriess, *The Desert Column* (Angus and Robertson, 1932), p. 54.

35. Quoted in *The History of the Seventh (Service) Battalion of the Royal Sussex Regiment 1914–19*, op. cit., p. 25.

36. William March, *Company K* (Gollancz, 1933), p. 151. March's book took the form of an experimental novel of 118 'mini-chapters', each headed by the name of a particular individual in the company, who told his story of a particular event.

37. McBride, p. 106.

38. Taylor, op. cit., pp. 83–4.

39. J.C. Dunn, *The War the Infantry Knew* (Cardinal/Sphere Books, 1989), p. 84.

40. *London Gazette*, 6 September 1917.

41. Taylor, op. cit., pp. 87–8.

42. Hesketh-Prichard, op. cit., p 82.

43. Hesketh-Prichard, op. cit., pp. 116–17.

Chapter Five
The Second World War

1. Wallace Reyburn, *Rehearsal for Invasion*.

2. Ronald Atkin, *Dieppe 1942* (Macmillan, 1980), pp. 101–2.

3. Quoted in Atkin, op. cit., pp. 102–3.

4. Vasili Ivanovich Chuikov, *The Beginning of the Road* (MacGibbon and Kee, 1963), pp. 141–5.

5. Erich Kern, *Dance of Death*, trans. Paul Findlay (Collins, 1951), pp. 179–80.

6. Captain C. Shore, *With British Snipers to the Reich* (Small-Arms Technical Publishing, 1948), pp. 128–9.

7. Shore, op. cit., pp. 135–6.

8. Ernie Pyle, *Brave Men* (Henry Holt, 1944), pp. 255–6.

9. Sergeant Harry Furness, note to author.

10. Shore, op. cit., pp. 63–5.

11. Barry Wynne, *The Sniper* (MacDonald, 1968), pp. 23–4.

12. Furness, op. cit.

13. Furness, op. cit.

14. Peter Young, *Storm from the Sea* (William Kimber, 1958), pp. 175–6.

15. Furness, op. cit.

16. Shore, op. cit., p. 75.

17. Furness, op. cit.

18. Shore, op. cit., pp. 76–7.

19. Charles Askins, *Unrepentant Sinner* (Paladin Press, 1985), pp. 148–9.

20. Askins, op. cit., pp. 150–51.

21. Shore, op. cit., pp. 84–5.

22. Richard Tregaskis, *Guadalcanal Diary* (Angus and Robertson, 1943), pp. 110–11.

23. Tregaskis, op. cit., p. 217.

24. US Intelligence Bulletin No. 3, November 1942, quoted in Peter R. Senich, *US Marine Corps Scout Snipers: World War II and Korea* (Paladin Press, 1993), pp. 182–3.

25. *Marine Corps Gazette*, January 1945, quoted in Senich, op. cit., p. 186.

26. Tregaskis, op. cit., pp. 133–4.

27. Russell Braddon, *The Naked Island* (T. Laurie, 1952), p. 77.

28. Hargis Westerfield, *The Jungleers: A History of the 41st Infantry Division* (41st Association, 1980), pp. 3–4.

29. Quoted by James A. Dunnigan, 'The American Sniper', in *Guns and Ammo Annual*, 1981, pp. 51–2.

30. Westerfield, op. cit., p. 4.

Chapter Six
Containing Communism

1. Lieutenant-Colonel Glen E. Martin, 'They Call Their Shots' in *Marine Corps Gazette*, April 1953.

2. Major Norman W. Hicks, 'Team Shots Can Kill' in *Marine Corps Gazette*, December 1963.

3. Gunnery Sergeant Francis H. Killeen, 'Sniping: Notes from a Career Marine' in N.A. Chandler and Roy Chandler, *Death from Afar*, vol. IV (Iron Brigade Armory, Maryland).

4. Killeen, op. cit.

5. Master Sergeant John E. Boitnott, 'Sniping: My Man Friday' in

N.A. Chandler and Roy Chandler, *Death from Afar*, vol. II (Iron Brigade Armory, Maryland), pp. 102–3.

6. Corporal Chet Hamilton, in Charles W. Sasser and Craig Roberts, *One Shot – One Kill* (Pocket Books, 1990), pp. 90–91.

7. Sergeant Tom Nowell MM, 1st Battalion Duke of Wellington's Regiment, account submitted to Nigel Greenaway and provided to author.

8. Ralph T. Walker, quoted in Peter Senich, *Limited War Sniping* (Paladin Press, 1977), pp. 17–18.

9. Walker, op. cit.

10. Master Sergeant William D. 'Bill' Abbot, 'Sniper Problems and Solutions, RVN 1966–1967' in N.A. Chandler and Roy Chandler, *Death from Afar*, vol. III, pp. 47–9.

11. Lieutenant-Colonel Thomas D. Ferran (ARNG), interview with author.

12. Ferran, op. cit.

13. Steve Suttles, 'A Sniper Then and A Sniper Now' in Chandler and Chandler, op. cit., vol. IV.

14. Sergeant Jeffrey L. Clifford, 'In His Own Words' in Chandler and Chandler, op. cit., vol. II, p. 44.

15. Clifford, op. cit. p. 45.

16. Clifford, op. cit., pp. 45–6.

17. Gunnery Sergeant Carlos N. Hathcock, from video recording *Carlos Hathcock: Marine Sniper* (L.O.T.I. Group in association with Paladin Press, 1994).

18. Jim Spencer, 'The Sniper' in the *Chicago Tribune*, 7 September 1986.

19. Thomas D. Ferran, account supplied to author.

20. Sergeant Bobby Sherrill, 'Sniping: Chinese Officer' in Chandler and Chandler, op. cit., vol. II, pp. 89–90.

21. The Overseas Weekly, 7 December 1968, quoted in Peter Senich, *The Long-Range War: Sniping in Vietnam* (Paladin Press, 1994), p. 34.

22. Major Powell, report, February 1969, quoted in Senich, op.cit., pp. 82–3.

23. 9th Infantry Division after-action report, 4 February 1969, quoted in Senich, op.cit., pp. 84–5.

24. Carlos Hathcock, op. cit.

25. Mark Limpic, 'Sniping: The Virtue of Patience' in Chandler and Chandler, op. cit., vol. IV.

26. Ferran, op. cit.

27. Ferran, op. cit.

28. Ferran, op. cit.

29. Craig Roberts and Charles W. Sasser, *The Walking Dead: A Marine's Story of Vietnam* (Grafton Books, 1990), pp. 200–1.

30. Ferran, op. cit.

Chapter Seven
Small Wars and the End of Empire

1. SAS soldier, quoted in *SAS: The Complete Record of the World's Leading Fighting Force* (Orbis Publishing/Brown Packaging, 1995), p. 144.

2. Guardsman Mike James, account sent to author.

3. Marine Mick Harrison, interview with author.

4. Harrison, op. cit.

5. David Cooper, interview with author.

6. Cooper, op. cit.

7. Cooper, op. cit.

8. Corporal Steve Newland, quoted in *War in Peace* (Orbis Publishing, 1985), p. 2235.

9. Cooper, op. cit.

10. Sergeant, Royal Green Jackets, quoted in Max Arthur, *Northern Ireland: Soldiers Talking* (Sidgwick & Jackson, 1987), p. 87.

11. David Barzilay, *The British Army in Ulster*, vol. 4 (Century Books, 1981), p. 191.

12. Officer, 1st Royal Anglians, quoted in Charles Allen, *The Savage Wars of Peace* (Michael Joseph, 1990), p. 219.

13. Sniper Officer, interview with author.

14. Royal Marine Sniper, 40 Commando, account submitted to Nigel Greenaway and provided to author.

15. Sniper Officer, op. cit.

16. Royal Marine Sniper, 40 Commando, op. cit.

17. Sniper Officer, op. cit.

18. Sniper Officer, op. cit.

19. Corporal, 2 Para, quoted in Arthur, op. cit., p. 185.

20. Sniper NCO, taped account sent to author.

21. Sniper NCO, op. cit.

22. Comment made to author.

23. Sniper NCO, op. cit.

24. Sniper Officer, op. cit.

Chapter Eight
Training the Sniper

1. Captain Tim Hunter, quoted in P.L. Thompson, 'Scout-Sniper School' in *Leatherneck*, March 1984.

2. Gunnery Sergeant Paul S. Herrmann, quoted in Thompson, op. cit.

3. Carlos Hathcock, quoted in Craig Roberts, 'Master Sniper's One Shot Saves' in *Soldier of Fortune*, May 1989.

4. Chuck Kramer, quoted in an interview with Dale A. Dye, 'Chuck Kramer: IDF's Master Sniper' in *Soldier of Fortune*, September 1985.

5. British Paratroop Officer, interview with author.

6. British Sniper Officer, interview with author.

7. John Coleman, 'One Shot – One Kill' in *Soldier of Fortune*, December 1986.

8. British Sniper NCO, account supplied to author.

9. British Paratroop Officer, op. cit.

10. Sergeant Brian Poor, quoted in 'Sniping – The Observer's Weapon' in N.A. Chandler and Roy Chandler, *Death from Afar*, vol. II (Iron Brigade Armory, Maryland), pp. 45–6.

11. Sergeant Gary Grella, in S/Sergeant Ernie Carter, 'Snipers Train to Stalk their Prey', in *Marines*, November 1988.

12. British Paratroop Officer, op. cit.

Chapter Nine
Sniping in Contemporary Conflicts

1. Lieutenant-Colonel R.D. Bean, quoted in N.A. Chandler and Roy Chandler, *Death from Afar*, vol. II (Iron Brigade Armory, Maryland) p. 60.

2. Corporal Tom Rutter, quoted in Charles W. Sasser and Craig Roberts, *One Shot – One Kill* (Pocket Books, 1990), pp. 234–6.

3. Craig Roberts, 'American Snipers in Beirut' in *Soldier of Fortune*, August 1989, pp. 79–80.

4. Chuck Kramer, quoted in an interview by Dale A. Dye, 'Chuck Kramer: IDF's Master Sniper' in *Soldier of Fortune*, September 1985, pp. 63–4.

5. Chris Hedges, 'War is Vivid in the Gun Sights of the Sniper', *The New York Times*, 3 February 1991.

6. Letter written to Jim O'Hern, and reproduced in N.A. Chandler and Roy Chandler, *Death from Afar*, vol. III (Iron Brigade Armory, Maryland), p. 158.

7. Account reported to Chandler and Chandler, and reproduced in op. cit., vol. II, p. 125.

8. Account in Chandler and Chandler, op. cit. vol. III, p. 162.

9. John L. Hogan, 'Snipers of Vinkovci' in *Soldier of Fortune*, March 1993, p. 40. For a similar view of Croatian shooting prowess, see also Bob Jordan, 'Looking For a Few "Pretty Good" Shots' in *Soldier of Fortune*, May 1993, pp. 46–7.

10. Hogan, op. cit., pp. 40–41.

11. Skippy Hampstead, interview with author.

12. Hampstead, op. cit.

13. Hampstead, op. cit.

14. Hampstead, op. cit.

15. Lee Marvin, from his unpublished ms, 'Bloodbath in Bosnia', supplied to author.

16. Robert C. Mackenzie, 'Shoot and Scoot' in *Soldier of Fortune*, October 1993, pp. 50–51.

17. Richard Lucas, 'Sarajevo's Streetsweepers' in *Soldier of Fortune*, February 1995, p. 46.
18. Lucas, op. cit., p. 47.
19. Quoted in Lucas, op. cit., pp. 47–8.

Postscript
At the Receiving End

1. George Orwell, Homage to Catalonia (Secker and Warburg, 1938), pp. 250–53.

Bibliographical Notes

General Works

An account of US sniping, complete with vivid eye-witness descriptions, can be found in *One Shot – One Kill* (Pocket Books, 1990) by Charles W. Sasser and Craig Roberts. Peter Senich has produced two illustrated volumes, *The Pictorial History of US Sniping* (Paladin Press, 1980) and *The Complete Book of US Sniping* (Paladin Press, 1988). On the British side there is Ian Skennerton's *The British Sniper: British & Commonwealth Sniping and Equipment 1915–1983* (Skennerton/Arms & Armour Press, 1984). A highly detailed analysis of sniping tactics and techniques – including diagrams and photographs – is explained in *The Ultimate Sniper* (Paladin Press, 1993) by John L. Plaster. My own *Sniper: One-on-One* (Sidgwick & Jackson, 1994) covers the development of sniping from the American Revolution to the present day and includes chapters on the training and the role of the modern sniper.

The many books devoted to sniper rifles include William H. Tantum, *Sniper Rifles of the Two World Wars* (Ottawa Museum Restoration Service, 1967), Duncan Long, *Modern Sniper Rifles* (Paladin Press, 1988) and the sniping chapter of Chris Bishop and Ian Drury (eds), *The Encyclopedia of World Military Weapons* (Temple Press, 1988/Crescent Books, 1989).

Chapter One
The Long Rifle and the American Revolution

The history and use of the American long rifle is discussed in the relevant chapters of Harold L. Peterson, *The Book of the Gun* (Golden Press/Paul Hamlyn, 1962/1963) and Harold L. Peterson and Robert Elman, *The Great Guns* (The Ridge Press/Hamlyn, 1971). Further detail is given in George C. Neumann, *The History of the Weapons of the American Revolution* (Bonanza Books, 1967) and Warren Moore, *Weapons of the American Revolution* (Promontory Press, 1967). Also useful, if less accessible, are Berkely L. Lewis, *Small Arms in the United States Service* (Smithsonian Institute, 1956) and Charles Winthrop Sawyer, *Firearms in American History* (privately published, 1910).

Firsthand accounts of the riflemen who fought in the American Revolution can be found in Henry Steele Commager and Richard B. Morris (eds), *The Spirit of the Seventy-Six* (Harper and Row, 1958). Other related works include Craig Boddington, *America: The Men and the Guns that Made her Great* (Peterson Publishing, 1981) and Roy F. Chandler, *Tim Murphy – Rifleman* (Bacon & Freeman Publishers, 1993), the latter a novel of Murphy's life and exploits.

Chapter Two
Sharpshooting in Europe

David Gates' *The British Light Infantry Arm c. 1790–1815* (Batsford, 1987) is a thoroughly researched work which contains much information on the development of rifle-armed troops in the British Army during the Napoleonic Wars. Although with a chronological range extending beyond this chapter, also of interest is Charles Chevenix Trench, *A History of Marksmanship* (Longmans, 1972).

Of the many accounts written by light infantrymen who fought in the Peninsular campaigns, among the more readable (and easy to find) are Edward Costello, *The Peninsular and Waterloo Campaigns*, ed.

Anthony Brett-James (Longmans, 1967), *The Recollections of Rifleman Harris*, ed. Christopher Hibbert (Leo Cooper, 1970), and *The Letters of Private Wheeler*, ed. Captain B.H. Liddell Hart (Michael Joseph, 1951).

Chapter Three
The American Civil War

Berdan's Sharpshooters are comprehensively dealt with in Roy M. Marcot, *Civil War Chief of Sharpshooters: Hiram Berdan — Military Commander and Firearms Inventor* (Northwood Heritage Press, 1989) and *Wiley Sword, Sharpshooter: Hiram Berdan, his Famous Sharpshooters and their Sharps Rifles* (Andrew Mowbray Inc., 1988). An original history of the regiment by Charles Stevens had recently been reprinted: *Berdan's United States Sharpshooters in the Army of the Potomac — 1861 to 1865* (Morningside Bookshop, 1984). Among the accounts of former Sharpshooters, of note are Rudolf Aschmann, *Memoirs of a Swiss Officer in the American Civil War*, ed. and int. Heinz K. Meier (Herbert Lang, 1982) and the *Civil War Diary of Wyman S. White* (unpublished, but extensively quoted in Marcot, above). Also of interest is *The Civil War Diary of Rice C. Bull, 123rd New York Volunteer Infantry*, ed. Jack Bauer (Presidio Press, 1977).

From the perspective of the South, an essential book is John Anderson Morrow, *The Confederate Whitworth Sharpshooters* (privately published, 1989), which also contains a full bibliography. William C. Davis has written a new history of the Orphan Brigade at war, *The Orphan Brigade: The Kentucky Confederates Who Couldn't Go Home* (Doubleday, 1980), which complements Ed Thompson's *History of the Orphan Brigade* (Press & Bindery of Chas. T. Dearing, 1898). Firsthand Orphan accounts include *Johny Green of the Orphan Brigade*, ed. A.D. Kirwan (University of Kentucky Press, 1956) and *Diary of a Confederate Soldier: John S. Jackman of the Orphan Brigade*, ed. William C. Davis (University of South Carolina Press, 1990). The many volumes of the *Confederate Veteran* contain a mine of information, for those with sufficient time and the access to this work.

The books by Marcot, Sword and Morrow (above) all contain information of the weapons used by sharpshooters in the Civil War; further details can be found in the relevant sections of William C. Davis, *Rebels & Yankees: The Fighting Men of the Civil War* (Salamander Books, 1989) and Ian V. Hogg, *Weapons of the American Civil War* (Bison Books, 1987).

Chapter Four
The First World War

Sniping in France: With Notes on the Scientific Training of Scouts, Observers and Snipers by Major H. Hesketh-Prichard remains one of the classic accounts of sniping in any war. A single edition was published by Hutchinson in 1920, making it a rarity until reprints in the US by Lancer Militaria (1993) and in Britain by Leo Cooper (1994). Other important works on sniping in the trenches are those of Herbert W. McBride, *A Rifleman Went to War* (Samworth Books, 1935, reprinted by Lancer Militaria, 1987) and three privately published volumes by Major F.M. Crum: *Scouts and Sniping in Trench Warfare; With Riflemen, Scouts and Snipers* and *Memoirs of a Rifleman Scout. The Story of the Lovat Scouts* (The Saint Andrew Press, 1981) by Michael Leslie Melville is particularly relevant for this period, although the book covers the Scouts in later conflicts.

Writing from an Australian viewpoint, Ion L. Idriess produced two volumes describing sniping at Gallipoli and in Palestine: *The Desert Column* (Angus and Robertson, 1932) and *The Australian Guerrilla – Book II: Sniping* (Angus and Robertson, 1942). Although N.A.D. Armstrong published his *Fieldcraft, Sniping and Intelligence* (Gale & Polden, 1942) during the Second World War for the training of sniper recruits, the content of the book is based on Armstrong's experiences as a sniper with the Canadian Corps in the Great War.

Many memoirs of the First World War refer to sniping; among the more famous are Robert Graves, *Goodbye to All That* (Cape, 1929), Siegfried Sassoon, *Memoirs of an Infantry Officer* (Faber, 1931),

Charles Edmonds (C.E. Carrington), *A Subaltern's War* (Peter Davies, 1929) and J.C. Dunn, *The War the Infantry Knew* (P.S. King, 1938) – modern editions ensure that all these volumes remain in print. Other books which make reference to sniping include Ian Hay, *The First Hundred Thousand* (Blackwood, 1916), Oliver Hogue, *Trooper Bluegum at the Dardanelles* (Melrose, 1916) and F.A.J. Taylor, *The Bottom of the Barrel* (Regency Press, 1928).

The novel by Frederick Sleath, *Sniper Jackson* (Herbert Jenkins, 1919), views sniping with an accuracy borne out of direct experience. Although a work of sociology, Tony Ashworth's *Trench Warfare 1914–18 – the Live and Let Live System* (Macmillan, 1980) uses a number of sniping incidents to support his argument. Stephen Bull's article, 'British Army Snipers, 1914–18', in *Military Illustrated*, September 1992, provides a short but comprehensive overview.

Chapter Five
The Second World War

A wide-ranging survey of Second World War sniping can be found in Captain Clifford Shore's *With British Snipers to the Reich*, published by Small-Arms Technical Publishing in 1948, and reissued in 1988 by Lancer Militaria. Despite its title, Shore's book covers sniping in other armies, and considers weapons and marksmanship in general. Peter Senich's *The German Sniper 1914–45* (Paladin Press, 1980) describes German sniper weapons and tactics, and reproduces an interview with three leading German snipers. *US Marine Corps Scout Snipers: World War II and Korea* (Paladin Press, 1993), also by Peter Senich, has much information on the Pacific campaign, as does *Death from Afar*, vol. II by N.A. and R.F. Chandler (Iron Brigade Armory, Maryland).

Firsthand accounts include Peter Young, *Storm from the Sea* (William Kimber, 1958), Charles Askins, *Unrepentant Sinner* (Paladin Press, 1985) and Richard Tregaskis, *Guadalcanal Diary* (Angus and Robertson, 1943). Barry Wynne's novel, *The Sniper* (MacDonald, 1968), is an account of the career of British sniper, Arthur Hare.

Chapter Six
Containing Communism

The four volumes of *Death from Afar: Marine Corps Sniping* (Iron Brigade Armory, Maryland) by Norman and Roy Chandler contain many valuable firsthand accounts, along with informed comment on sniping in Korea and Vietnam. Also useful are Peter Senich's *Limited War Sniping* (Paladin Press, 1977), *US Marine Corps Scout Snipers: World War II and Korea* (Paladin Press, 1993) and *The Long Range War: Sniping in Vietnam* (Paladin Press, 1994).

Charles Henderson's *Marine Sniper: 93 Confirmed Kills* (Berkeley, 1988), is a biography of Carlos Hathcock with special emphasis on his experiences in Vietnam. *The Walking Dead* (Grafton, 1990) by Craig Roberts and Charles W. Sasser is the story of Craig Roberts' Vietnam tour, which included time as a Marine sniper. Joseph T. Ward's *Dear Mom: A Sniper's Vietnam* (Ivy Books, 1991) describes, among other things, the activities of top sniper Chuck Mawhiney, also the subject of a forthcoming book by Peter Senich.

Although not primarily concerned with sniping, information on US Army sniping in Vietnam is contained in Julian J. Ewell and Ira A. Hunt, *Sharpening the Combat Edge: The Use of Analysis to Reinforce Military Judgement* (Vietnam Studies, Department of the Army, 1974) and John H. Hay Jnr, *Tactical and Material Innovations* (Vietnam Studies, Department of the Army, 1974).

Chapter Seven
Small Wars and the End of Empire

A history of British sniping since 1945 has yet to be written, but a number of books are of interest. Charles Allen's *The Savage Wars of Peace* (Michael Joseph, 1990) provides an excellent overview of Britain's small wars from the soldier's viewpoint. On Northern Ireland there is Max Arthur, *Northern Ireland: Soldiers Talking* (Sidgwick & Jackson, 1987) and the four volumes by David Barzilay,

The British Army in Ulster (Century Books). For the Falklands, Martin Middlebrook's *Operation Corporate: The Falklands War* (Viking, 1985) has assembled many eye-witness accounts, some of which mention sniping.

Chapter Eight
Training the Sniper

John L. Plaster's *The Ultimate Sniper* (Paladin Press, 1993) has much useful information about sniping techniques. The two articles by Mick McIntyre, based on Royal Marine training, in *Combat and Survival* magazine (July and August 1992) provide a summary of British sniper techniques. A transatlantic approach is adopted in the two chapters from *Combat* (Marshall Cavendish, 1988), edited by Len Cacutt.

Declassified sniper manuals are becoming increasingly available and provide the reader with the 'nuts and bolts' of sniper training. Among these are the *SEAL Sniper Training Program* (Paladin Press reprint), *USMC Sniping manual* (Desert Publications reprint) and the *US Army Sniper Training Manual* TC-23-14.

Chapter Nine
Sniping in Contemporary Conflicts

The *Death from Afar* volumes by the Chandlers are again invaluable, with coverage of sniping in Grenada, Somalia and the Lebanon. The magazine *Soldier of Fortune* devotes much attention to contemporary conflicts, which include reports from its journalists on sniping across the globe. *Soldier of Fortune* also produces regular features on specific sniper weapon systems.

Index of Contributors